# Of Ants and Men

David G. Green

# Of Ants and Men

## The Unexpected Side Effects of Complexity in Society

Copernicus Books is a brand of Springer

David G. Green
Faculty of Information Technology
Monash University
Clayton, VIC
Australia

ISBN 978-3-642-55229-8      ISBN 978-3-642-55230-4   (eBook)
DOI 10.1007/978-3-642-55230-4
Springer Heidelberg New York Dordrecht London

Library of Congress Control Number: 2014937524

Printed on acid-free paper

Copernicus Books is a brand of Springer
Springer is part of Springer Science+Business Media (www.springer.com)
Copernicus Books
Springer Science+Business Media
233 Spring Street
New York, NY 10013
www.springer.com

# Preface

This book arose out of a previous book, *The Serendipity Machine*, in which I explored the many ways in which modern computing exploits unexpected discovery. The final chapter of that book briefly explored the ways in which the information revolution was itself leading society into unexpected "discoveries" about new ways of working and living. In the course of writing the book, it became evident that the unexpected consequences of using computers were symptomatic of a much wider set of processes at work in society. Exploring and developing this insight led directly to this book.

Before I am attacked for sexism, I should explain the title. The title is an allusion to the poem *To a Mouse*, written by Robert Burns in 1785. It contains the famous line:

> The best-laid schemes o' mice an' men gang aft agley.

Here I have adapted this line to refer to the way humans plans go wrong because we are like ants: busy with our individual concerns without being aware of the wider consequences.

Recently, a number of ideas about social complexity have surfaced in popular literature. Some have even entered the language as common expressions, such as "six degrees of separation" and "tipping point." However, for the most part they have been presented as interesting, but isolated ideas. One goal of this book has been to show how these ideas all fit into a larger, more comprehensive framework of social complexity. Networks provide a good model for understanding complexity. Here I use them to show how many familiar ideas arise as natural consequences of social and other networks.

The problem in trying to write about social complexity is that there is so much of it. The curse of trying to pull together so many ideas and issues into a coherent theory is that it is impossible to deal with any single issue in-depth. This applies both to individual examples, such as the origins of World War I, and to major issues, such as the discussion of environment in Chap. 19. It also applies to the geographic spread of examples. The risk is that by delving too deeply into such issues, the main themes could be lost among the detail.

Likewise, I have tried to provide examples, as well as statistical evidence, from many countries. However, in today's globalized world, similar issues tend to arise in every part of the developed world, so I have sometimes used cases from my home country Australia as typical examples. Overall I have tried to strike a balance between maintaining the thread of ideas and providing enough stories and examples to clarify them.

In all of the above cases, I have had to strike a balance between telling the whole story and clearly describing the role played by unconscious side effects in creating the situation. The problem is that many of the topics I need to touch on could be the subject of entire books; more than a few have been. Chapter 18, for instance, concerns economic growth, which has been the subject of countless books and debates. Likewise Chap. 19, which concerns the global environment, draws heavily on my book *Complexity in Landscape Ecology*, as do several examples presented in other chapters. Humanity's impact upon the environment has been the subject of intense debate and conflict. In both chapters, I have had to restrict myself to showing that these issues, of intense public debate, arise as unintended trends produced by human activity.

A lot went into the making of this book. It rests on years of observations. These have been filtered by theory, and tens of thousands of words have been deleted from early drafts.

I am grateful for the generous help and support of family, friends, and colleagues at every stage of this work. Several family members, Dr. Tony Green, Cheryl Green, and my wife Yvonne Green, provided many useful insights and suggestions. My daughter Hilary Green provided a cover picture. Colleagues also made significant contributions. I am grateful to Dr. Suzanne Sadedin and Les Hughes for many helpful discussions, as well as constructive feedback and comments on drafts of several chapters. Dr. Joanne Evans and Prof. Julie Fisher also provided useful feedback about several chapters. Dr. David Squire checked the entire text with his text analysis program Damocles. His cross chapter analysis helped me to avoid duplication and to build coherent threads of ideas and examples. His plagiarism check revealed just how easy it is to fall into using clichéd phrases! Marc Cheong provided useful feedback on drafts of the entire manuscript. Finally, Jacob Matthews did a thorough job of proof reading the manuscript, as well as providing many useful suggestions. I am grateful to all of the above for their help, support, and encouragement.

Melbourne, Australia                                                        David G. Green

# Contents

# From Bad to Worse

**1**

*Left to themselves, things tend to go from bad to worse.*[1]

## Abstract

Unexpected trends, events and even disasters happen all too frequently in modern society. These events range from problems facing us as individuals to national disasters. They are symptoms of the complexity of modern society. In many cases they arise as unintended side effects of every day activity. Complexity means that in any complex situation, any decision will have unintended consequences. Examples include immediate needs driving out important ones, cascading problems, and episodes of anarchy such as the New York 1977 blackout.

## A Night of Terror

*July 13, 1977.* A hot summer's evening in New York. Air conditioners are running at full speed. From the *Windows on the World* restaurant, high atop the World Trade Centre, diners look out over a sea of light that stretches to the horizon.

The first hint of the trouble to come is the arrival of a fierce thunderstorm.[2] It strikes early in the evening. High winds and rain move across Westchester County. At about 8:37 p.m., lightning strikes a power transmission line. Less than 20 min later, lightning strikes another section of line. The grid tries to cope with these breaks by sucking power from other areas. However, the increased demand soon overloads relays, which operators at once shut down. A chain reaction ensues. Relays are going down all over the city. By 9:21 p.m., all links to power sources outside the city have been cut. There is no more power. New York City is plunged into darkness.

D. G. Green, *Of Ants and Men*, DOI: 10.1007/978-3-642-55230-4_1,
© Springer-Verlag Berlin Heidelberg 2014

Meanwhile back at the World Trade Centre, a group of diners finish their meal. They step into the elevator. The doors close. Nothing happens. Annoyance turns to fear as they press buttons again and again with no result. For several minutes, several terrifying minutes, they remain suspended, one hundred and seven floors above the ground. In desperation they manage to force the doors open and scramble back into the safety of the now-darkened restaurant. An eerie blanket of pitch black has replaced the brilliant sea of light they had admired just minutes before.[3]

So began the infamous 1977 New York blackout. The ordeal lasted all night and most of the next day. For many inhabitants, especially in the Bronx, it was a night full of terror.

> Minutes after the Blackout began, men in trucks equipped with chains and hooks were being paid by crowds to rip off the iron gates and fences that protected neighbourhood stores. Within fifteen minutes, stolen goods were being offered to neighbourhood residents who were on the streets or stranded in apartment buildings without elevator service. Within two hours it became apparent that the situation was not going to end quickly, and thousands of otherwise law-abiding citizens joined in what was to become the largest collective theft in history.[4]

The frightening aspect of this report is that it shows just how easy it is for prevailing law and order to break down. The risk of such a breakdown is increased by the fragility of urban infrastructure, which makes it possible for unpredictable events to disrupt key services and unleash social chaos. As one report commented:

> It is challenging to imagine the blackouts as anything but technological accidents— oversights of management and engineering quickly and inevitably corrected. Yet these failures were far from random. The blackouts exposed to public view an intersection of long-term developments which have often been overlooked, underestimated, or studied as disconnected phenomena.[4]

The effects of the 1977 blackout were not confined to looting. Vandalism and arson were rampant. Police made 3,776 arrests during the blackout, but thousands more evaded capture. The Fire Department responded to some 1,037 fires, about six times the average, as well as another 1,700 false alarms.[2]

## From Bad to Worse

In the 1977 New York blackout, we see an example of how one thing leads to another, steadily getting worse all the time. One power station goes down and then another, each putting more pressure on the remaining stations until they all go down. This engineering problem, the blackout, then leads to another, even worse problem, as social chaos ensues.

Civilization is a delicate veneer. Like so many events in modern times, the New York blackout of 1977 shows how easy it is to shatter the illusion of control and plunge the world into chaos. Each incident creates waves that can spread out and wreak havoc on a much larger scale. What starts off bad, quickly grows worse. The effects can be totally unexpected, totally unpredictable.

The history of human civilization is, in large measure, a story of the human quest for control. After thousands of years of civilization, we think that we control the environment in which we live. We begin to think that we control the natural world. We might even fool ourselves into thinking that we control human nature. Modern society is built on the assumption of control. Yet, as the terror of the New York blackout shows, chaos all too easily bursts forth, reminding us how flimsy the illusion of control really is.

The root cause of much of the chaos that besets us is complexity, sheer complexity. From complex webs of interactions, chaos emerges. It is complexity that leads to unexpected problems, that turns order into chaos. As much as anything, the New York blackout, like most accidents and breakdowns, was a result of complexity. The power system did have backups and safeguards built in. But no one had anticipated that the network could suffer a cascade of failures of the kind that occurred. Nor could anyone anticipate the mayhem that would ensue when power failed on such a large scale for such a long period. This does not mean that the planners were incompetent; there are just so many possible ways that the system could behave, it is not possible to anticipate and plan for all contingencies.

Twenty-four years after the Night of Terror, New York suffered an even worse disaster. On September 11, 2001, two planes hijacked by terrorists crashed into the World Trade Center, bringing both towers crashing to the ground. Over 3,000 people were killed. The repercussions of that event spread around the world, affecting travel, commerce and behaviour in many countries. With people scared to fly, many airlines were driven to the point of bankruptcy. The tourism industry worldwide suffered a sudden, catastrophic decline.

## Unintended Consequences

The problem that plans and actions have unexpected outcomes has been a long-running concern of philosophers, economists and other thinkers. In 1936, the American sociologist Richard Merton proposed his *Law of Unintended Consequences*.[5] This law holds that in complex situations, any action you take leads inevitably to unanticipated outcomes. He suggested five underlying reasons for this:

1. *Ignorance*—people are never aware of all the factors, issues or processes involved in a highly complex environment.
2. *Error*—mistakes, including perpetuation of habitual ways of acting.
3. *Imperious immediacy of interest*—acting on the basis of immediate, local priorities or self-interest, rather than those of the entire group or society. One example is the *tragedy of the commons*.[6] This happens when a group shares a common resource. Although each individual acts responsibly in dealing with their portion, they do not consider the whole. As a result, the group ends up exhausting the entire resource.
4. *Values*—people often act according to social, traditional or other values, even if the consequences are contrary to best interests.

5. *Self-defeating prophecy*[7]—Sometimes fear drives people to find a solution to problems that never actually occur. For instance, fearing that a partner will try to take over his business, a manager cuts the partner out of vital meetings and spreads stories that he is trying to ruin the business. In reality the partner has no intention of taking over, but isolating him leads to missed opportunities and eventually the business fails.

Unintended consequences occur in many guises. We shall see many examples of this in later chapters. Unintended consequences can have both positive and negative effects. For example, the Unites States introduced prohibition during the 1920s as a way to reduce the social evils associated with alcohol consumption. However, it quickly led to an illegal trade in alcohol and allowed organized crime to take control of the industry. A similar problem has plagued more recent efforts to suppress drug use by legal means.

It is well-known in medicine that new drugs will have unexpected side effects. Some drugs have negative effects, such as cancer treatments that cause patients to lose their hair. But sometimes the effects are positive. Aspirin, for example, was developed originally for pain relief. However it was subsequently found to act as an anticoagulant, which led to its use as a treatment to reduce the risk of heart attacks and strokes.

The case of aspirin is an example of serendipity, accidental discovery, which has always played an important role in science. In a previous book, I described the many ways in which the computer acts as a sort of *serendipity machine*, allowing science to use accidental discovery deliberately, as a routine tool.[8]

Merton's famous study treated the problem of social complexity from the perspective of decision-making, especially decision-making by governments. However, his rules also help to explain some of the ways in which our day-to-day activities produce unexpected social outcomes. In particular, Rule 3 says that unanticipated outcomes emerge when people behave according to local self-interest, oblivious to (or ignoring) the wider implications of their actions.

Beginning in the late twentieth century, research has discovered much about the nature of complexity, both in nature and in human society. In later chapters we will see that one of the root causes of complexity lies in the networks formed by relationships and interactions. We will see that complex networks are embedded in virtually every aspect of society. They impinge on many social phenomena. In particular, they underlie most of Merton's rules, and especially Rule 3. But first we will look at a few examples that reveal some of the above rules in action.

## The Immediate Drives Out the Important

*8 p.m., Thursday night. Women's dorm. Chaos College.*
Lindy is sitting at her desk poring over textbooks and chewing her nails.

"Hey Lindy, we're all going out to celebrate the end of the working week a day early. Wanna come?"

"Sorry, Meg. This assignment is due in tomorrow morning. I've got to get it finished."

"Huh, that's short notice, isn't it?"

"Actually, they gave it to us three weeks ago."

"Oh. But surely, you must be nearly finished. Why not knock it off in the morning?"

"Nah, I can't. It's way too big. And I've only just started. I'll be lucky to get any sleep tonight."

"How come?"

"I just haven't had time."

"Why?"

"Well, you know. First, there was that dinner we organized. Then Sue needed help moving into her new apartment. And there was a lot of reading for my other classes. I've been swamped."

Most of us have been caught in this kind of predicament. If it is not an assignment, then maybe it is a report, a brief, a manuscript, or a project. Really, it could be any activity that is important to our future and has a deadline. Just like Lindy, there are often many other immediate pressures that distract us from getting the important things done.

One common example of the immediate driving out the important occurs in politics. In democratic countries, politicians are notoriously obsessed with getting themselves re-elected. Political self-interest need not be blatant, or calculating. Politicians are simply slaves to their professional obsessions, the first of which is winning the next election. This obsession leads them to favour policies with immediate visible results over long-term projects in the best interests of the country. If the interval between elections is (say) 3 or 4 years, then that is their planning horizon. Thus they will tend to favour quick fixes over permanent solutions. They will put important, but long-term issues on the back burner. It is only when authorities have secure, long-term tenure that they are likely to consider long-term matters.

In 2002, the tendency to think about problems in terms of local cause and effect led to a crisis in Australia's most important river system, the Murray River. When combined with its 41 tributaries, many of them important rivers in their own right, the Murray's catchment encompasses 14 % of the continent's land area, and is crucial to much of the country's agricultural production. Its historical, social, environmental and economic significance prompted the National Trust to declare it a heritage site.

Given its economic importance, farmers, councils and government agencies have always seen the need to solve local environmental challenges associated with the river. Despite these good intentions, their solutions usually reflected local thinking, and did not take into account the health of the whole river system. Sometimes they simply moved pollution and other problems from one location to another. In 2002 the National Trust of Australia was forced to declare the river an endangered heritage site: "Today no water flows into the sea from the Murray River. This once magnificent river now regularly fails to reach the sea."

The significant threats to the health of the Murray remain largely unaltered. However, the community engagement process, along with the work of community organisations has raised the profile of the Murray issue considerably. National action is required, and the

National Trust urges the Council of Australian Governments to commit to a national
approach to ensure that healthy flows are restored to the Murray, and that the community
is fully informed and engaged in the conservation of our most precious resource—water.[9]

The problem revealed itself as many separate issues. Diverting water from the
rivers for irrigations led to floodplains and wetlands drying up. In other places,
floodplains became permanently inundated where dams held back water. This in
turn led to greater pollution and sedimentation of the Murray itself. Other problems included declining rainfall, salinization, desertification, and the introduction
of European carp and other exotic species. By the turn of the millennium, all these
changes were having severe ecological consequences. Native fish, invertebrates
and reptiles were in decline and several species of waterbird became extinct in the
region.[10]

## The Challenge of Complexity

Complexity is a problem that has dogged civilization since the dawn of time. Most
of the institutions and methods that people use to control it have evolved by simple
trial and error over long periods of time. Only in recent times have people consciously set out to understand the nature of complexity. The world today is
exceedingly complex. If we are to cope with that complexity, both as individuals
and as societies, then we cannot afford to ignore it.

Complexity is the richness in structure and behaviour that emerges from
interactions within a large system. A school of fish emerges when lots of individual
fishes coordinate their behaviour and swim together. The development of powerful
computers during the late twentieth century made it possible to study complexity
and much is now known. We know, for instance, that underlying all complex
phenomena are networks of connections and interactions. We know that complex
systems are inherently unpredictable. We know that interactions form feedback
loops that can keep a system stable, or else drive it to spiral out of control.[11]

In the years immediately leading up to the year 2000, the media made a huge
fuss over the Millennium Bug, a fatal flaw at the heart of computers that would
trigger the collapse of the entire socioeconomic framework. Or so the most sensational reports would have the public believe. In the event, there was no Y2K
disaster. Simple fixes avoided problems on older machines, and computers
everywhere turned over to 1/1/00 without the world crashing to a halt.

However, what was generally overlooked in the debate was that the turn of this
millennium really does represent an unprecedented turning point in all human
history. One of the greatest challenges that the world faces at the start of the third
millennium is complexity. Never before in history have people faced such vast
changes as we do today. Never before in history have technological changes come
in such profusion, at such speed, and with such vast, entangled consequences.
Everything seems to be boiling up at once. By the beginning of the year 2000, the
world's population had reached unprecedented levels—nearly four times what it

was in 1900—and is still growing unchecked. In the space of a single lifetime, vast changes have taken place both in our social fabric and in the world's political order. Meanwhile, species are going extinct at the fastest rate since the demise of the dinosaurs, 65 million years ago.

In the face of such enormous upheavals, we really are justified in entering the new millennium with a sense of trepidation. Somewhere, somehow, something has to give. All at once, we are faced with the very real prospect of blowing ourselves up with nuclear weapons; of poisoning the environment so that it becomes uninhabitable; of dying horribly in a pandemic unleashed by genetic meddling; of dying of thirst in a runaway greenhouse effect; or of being butchered in some violent antisocial meltdown. Perhaps we really are witnessing the apocalypse unfolding.

Not only is each of the above changes a cause for concern in itself, but also the interplay between them. For example, population growth leads to environmental hazards, but the attempts to remove the hazards often run into social, political and economic barriers.

Industrial society has become highly dependent on centralized systems and services. Two events during the year 1998 serve to highlight how fragile our big cities really are. First, the water supply of Sydney became contaminated with algae. Besides making many people ill, it also led to a desperate water shortage that lasted for many weeks.

Later in the same year a gas storage tank exploded on the outskirts of my home town Melbourne. Besides killing several people, the explosion also cut off the supply of natural gas to the entire state of Victoria for weeks. The blowout forced millions of people to take cold showers, disrupted business, put thousands out of work and cost the community hundreds of millions of dollars.

The modern world is increasingly complex and intertwined. What are seemingly innocuous changes in one sphere of activity can lead to catastrophe elsewhere. If we—the human race—are going to survive and flourish, then we need to learn how to cope with that complexity.

In the next few chapters we will look in more detail at ways in which disasters and other unexpected consequences emerge as a result of complexity. Then in later chapters we will look at how these processes contribute to several important trends in society as well as their implications.

## The Best Laid Plans

Despite our best intentions, despite all our planning, things often go badly wrong. And not only do they go wrong, they often go from bad to worse. Sometimes it is our very attempts to fix a problem that make things get worse, as they did for poor Harry in the story that follows.

Harry's plan was simple.[12] Wind up his business early, fly home and surprise his wife on her birthday. His meetings were due to finish next morning, but things were going well. He could move tomorrow's meetings forward to this afternoon, jump on the evening shuttle flight and be back home in time for his wife's birthday dinner. She had been disappointed, not to mention angry, when he told her he had to be away on business. They had planned it for weeks. Many of her friends and family would be there.

As he dialled for a taxi, Harry thought what a surprise it would be for her. His presence would be a gift in itself.

Then his problems began. The taxi took half an hour to arrive. As they drove off the driver explained that there were often delays late in the afternoon because, like Harry, lots of people wanted to leave for the airport. And besides, there was a convention on, so lots of people were arriving or leaving.

Outside the taxi it was raining and the peak hour city traffic seemed to grind to a standstill. Soon they were caught in a traffic jam. The street seemed more like a giant car park than a major thoroughfare.

Harry kept looking at his watch. The minute hand seemed to be moving faster than his taxi! Soon it was clear that he would not make the flight in time.

Better late than never, Harry thought. Whipping out his phone, Harry called the airline. By some miracle the next flight was not full and he was able to switch his booking. Relieved, he sank back into his seat. Now he had plenty of time. Feeling more relaxed, he looked out the window taking in the passing crowds, vehicles and buildings.

When they arrived, the airport was a crowded blur of panic as passengers raced to catch their planes. Having no luggage, Harry was able to skip the long queues at check-in and went straight to his gate.

He waited an hour at the gate along with a crowd of fellow passengers, but as boarding time approached, the notice board flickered and he was annoyed to see that departure of his flight was set back by 15 min. A tinny voice over the PA system announced that "due to late arrival of the aircraft, the flight will be delayed a few minutes."

Thirty minutes later the notice flickered again and another voice announced that engineers were working to correct a fault with the hydraulic system. Harry started to get worried. He did some quick mental calculations. There was one later flight, the last one for the evening. If he caught that flight, he would arrive home some time after 10 p.m. He'd be too late for the dinner, but still in good time to join the party and surprise his wife.

He approached the desk to ask what was happening. Several other passengers were already leaving. When one of them asked about switching to the late flight, Harry did the same.

"Sorry, but the late flight is already full. We're very busy this evening," the attendant informed him.

An hour came and went. With more than a little annoyance, Harry heard the announcement for the late flight to board. If he had booked on that flight first up, he would be on his way now. But still he waited.

Another voice announced that due to faults in the aircraft, it would be replaced. They were just waiting for the replacement aircraft to arrive and be serviced, and then they would be ready to board.

Half an hour later, there was still no sign of their plane. At last, the weary passengers watched in relief as the new plane taxied up to the gate and disgorged its passengers into the terminal. Harry, who had spread out papers over two seats and a table, gathered his hand luggage together ready to board.

"We regret that owing to late arrival of the aircraft, it will not be possible for the outbound flight to reach its destination before the airport curfew at 11 p.m. The flight is therefore cancelled."

Harry was devastated. His plan to surprise his wife was in ruins. Not only that, he had cancelled his hotel booking for the night. There was a convention in town and the hotels were all booked solid.

Harry's predicament highlights the way events can cascade out of control. In later chapters we will look in more detail at ways in which this occurs.

---

## End Notes

[1]Corollary to Murphy's Law: "whatever can go wrong, will go wrong."

[2]The account given in this section is based on details provided by the *New York Blackout History Project*.

[3]Reported in Newsweek July 25, 1977.

[4]Eyewitness report from the *New York Blackout History Project*.

[5]Merton (1936).

[6]Hardin (1968).

[7]Note that it is a self-defeating prophecy, not a self-fulfilling one. The prophecy does not necessarily come true (i.e. self-fulfilling) but does lead to unwanted problems (i.e. self-defeating).

[8]Green (2004).

[9]National Trust of Australia (2002). *Endangered Places—2002 Report Card*. Canberra, National Trust of Australia.

[10]Kingsford (2000).

[11]Such loops are called feedback. Negative feedback dampens changes and keeps a system stable. A thermostat, for instance, turns off power to an oven when the temperature becomes too high. In contrast, positive feedback reinforces changes, becoming more and more extreme. An example is compound interest, which makes sums of money grow at an ever-increasing rate. Both kinds of feedback play important roles in complex systems. See Chap. 10 for a discussion.

[12]This story, like many anecdotes in this book, is fiction, but based on an amalgam of actual events.

# Of Ants and Men

**2**

*The best laid plans of mice and men gang aft agley.*[1]

**Abstract**

People learn to respond to the world by acquiring and adapting schemas—recipes for dealing with particular situations. Such recipes can lead people to fall into habitual ways of thinking and acting. Habits make it possible for side effects of our activity to emerge without our being aware of them. This aspect of human behaviour makes us like the ants. One result is that a kind of natural selection works on people, creating unexpected social trends and patterns. Just as ants build their nests without conscious planning, so too social trends and patterns emerge as unexpected side effects of human activity. Examples include left-handers being good at certain sports and escalating differences in real estate prices.

## Why are Left-Handers Good at Sport?

*Sunday 6 July 2008.* It is the climax of the 2008 Wimbledon tennis championship: the men's singles final. Out on centre court, five times winner Roger Federer is playing his twenty-two year old Spanish rival Rafael Nadal. After five sets spanning 4 h and 48 min, and two long rain delays, it is finally match point. Nadal serves and Federer's return hits the net. Suddenly it's all over. Nadal is the new Wimbledon champion. Another left-handed player has joined the ranks of champions.

Why are left-handers so good at sport? They account for less than 10 % of the general population and yet at the highest levels of tennis they make up over 20 % of top ranked players. Many champions have been left-handers, including Rafael Nadal and Rod Laver, arguably the best men's singles player of all time.

D. G. Green, *Of Ants and Men*, DOI: 10.1007/978-3-642-55230-4_2,
© Springer-Verlag Berlin Heidelberg 2014

The prominence of left-handers in sport is not confined to tennis. Left-handers perform equally well, if not even better, in many sports. The list is impressive: cricket, fencing, squash, to name just a few. In boxing, for instance, the "southpaw", the left-hander, is famous for being difficult to beat.

Why is this so? What is it about left-handers that makes them better at sport than the right-handed majority of the population? Many theories have been proposed, but the truth is simple. Ironically it stems from the very fact that left-handers are rare in the general population. Because left-handers are rare, right-handed beginners do not encounter them very often. When they do meet left-handers, they do not know how to cope. They are all at sea. On the other hand, the left-hander knows exactly what to do. Moreover he or she quickly learns how to exploit the right-hander's confusion. Of course left-handers sometimes encounter other left-handers. One of the funniest sights in sport is to watch two left-handed beginners trying to compete against one another.

In sports such as tennis, left-handed beginners are able to exploit the confusion of right-handed beginners. They tend to win often. For a beginner in sport there is no greater incentive to continue than winning. If you win you feel on top of the world. If you lose, then your interest in a new sport quickly wanes. So left-handers get a lot of positive reinforcement, but most right-handers get negative reinforcement. The result is that a very high percentage of left-handers are encouraged to continue in their chosen sport. In a very real sense they have a selective advantage over their right-handed contemporaries.

As they progress up the ladder of success and skill, the left-hander continues to exploit the right-hander's lack of experience. The proportion of left-handers in the more advanced levels of the sport continues to climb. Finally at the pinnacle of international competition the proportion of left-hander players has risen to 20 %. They no longer have the advantage of rarity; only native skill matters. However anyone at the peak of a sport is very definitely a winner anyway.

The above is not the full story. Not by any means. Left-handers do not have an initial advantage in all sports. It matters only in sports where the competitors face each other one on one. This includes most combat sports, such as boxing, fencing, and judo. It also includes many court games such as tennis, badminton, and squash. Some team sports, such as baseball and cricket, also have continual face offs. On the other hand there are many sports where being left or right-handed is irrelevant. Such sports include athletics, equestrian, gymnastics, shooting, sailing, and swimming.

The example of left-handers in sport highlights the way that unseen, unplanned processes shape human affairs, without people being aware of them. To summarize, we have now seen several important points. First there is a selective process that operates to weed out players in sports. Secondly left-handers have a selective advantage in many sports. Finally humans are subject to selective processes. What about other physical attributes? Do they affect performance in sports as well? The answer of course is yes! To begin with, everyone knows that particular sports favour players with particular physiques: basketball players need to be tall, gymnasts should be light but muscular, and long distance runners need long legs and a light build. And so it goes on.

## Location, Location, Location

The success of left-handers in sport is just one example of unplanned outcomes that emerge from processes operating in human society. Take the way shops are distributed around a town. A few years ago, the video store down the road from our house closed and moved downtown. The reason for the move was that another video store had opened downtown and was already operating successfully. Businesses of a particular type tend to congregate together in the same part of town. This is not necessarily because those areas have been designated for (say) theatres or law firms. Rather it is because no one wants their competitors to gain an advantage over them. If you set up your business near a competitor, you avoid losing any advantage that their location gives them. You also give yourself the chance to steal customers away from them. As more theatres cluster together, the area becomes known as the theatre district. It is then imperative for new theatres to open in the same area or face a potential loss of patrons.

The same process also operates on other aspects of a city as well. Although most cities in the world today have central planning authorities, most of the older ones grew up in more or less haphazard fashion. This allows business districts and other concentrations to form. Even in the most highly planned of modern cities; the same effect is still at work. The real estate industry sees to that.

In the real estate industry, location is all-important in determining the market value of properties. A good house in a slum district will not fetch a high price, no matter how good it is. But how is it that some locations come to be better than others? Usually, it starts out with a natural advantage. One area may be conveniently placed near to major businesses, or it may be close the sea or a river, or it may be slightly hilly, allowing good views. These natural advantages are enough that people will seek them out and pay slightly more for them than they would for other properties. When some properties attract elevated prices, they raise the prices of other properties nearby. In this way, the average prices in one area will drift to become higher than in neighbouring areas. People naturally assume that the area with higher prices must be better to live in. So the process escalates. Over time, some districts become prestigious and much sought after, while others go into decline as people try to move to better areas. This is an example of the "snowball effect", which we will look at in Chap. 10.

## The Ant Hill

At first sight it looks like nothing more than a large, flattened mound of coarse, dirty sand under the trees. Then you notice ants swarming all over it and realize that you are looking at an ant colony. A large colony, it is at least 10 m in diameter. Looking more closely, you begin to make out dozens of tiny entrances, each guarded by soldier ants, with lines of worker ants passing to and fro. But what you see on the surface is only the beginning. Underground, it is a city in miniature.

Descending from the surface, a worker ant enters a vast network of passageways and chambers. It is a miracle of organization. There are storage chambers for food, nurseries for eggs, and even a hall where the queen ant holds court.

Few groups of animals, other than humans, create large societies with elaborate living quarters. Ants, bees, and termites are the best known, but they include several other groups of insects, and even some mammals, such as moles.

What we humans find most remarkable is that ants, bees and other insects, manage to create complex communities without the ability to design the structures they build. Ants are incapable of abstract thought. They do not consciously plan. And yet they manage to build highly organized colonies comparable to human cities in their complexity and structure.

The ability of ants to build miniature cities without deliberate planning and control holds many lessons. Perhaps the most important lesson of all is that the same kinds of unconscious processes are also at work in human societies.

We humans think that we are superior to the ants. We think that we plan and act rationally. We think that we are in control. And yet in many ways we are just like ants, busy doing everyday things, oblivious of the bigger picture, oblivious of the effects that our activities have. Just as ants create an ant hill without knowing what they are doing, so too humans create situations, social trends and problems without being aware of where our actions are leading.

The results of unconscious processes are all around us. They affect our daily lives. Often they shape and influence human society. Sometimes the results are undesirable, sometimes disastrous. Like the ants, we humans are only dimly aware of them. And yet we need to understand these processes. We need to understand why they occur. We need to understand how they occur. A good way to begin is to look more closely at what the ants do.

## Of Ants and Men

An ant has no idea of how to build an ant colony. All it knows are some simple rules that tell it how to behave. We might express some of these rules like this:

> *If you find a scrap of food, pick it up.*
> *If you find a pile of food, drop any scrap of food you are carrying.*

The effect of these rules is that the ants bring isolated scraps of food together (Fig. 2.1).[2] With hundreds of ants swarming around the ant hill, all behaving the same way, piles of food appear. Once the ants start forming these food piles, a process of positive *feedback*[3] comes into play. As ants continue moving scraps of food around, larger piles grow at the expense of smaller piles. Before long, the food becomes sorted into just a few large piles.

The above process, known as *stigmergy*,[4] is typical of the way in which order emerges spontaneously in many natural systems by self-organization. It is also an example of what has come to be known as *swarm intelligence*,[5] which refers to a

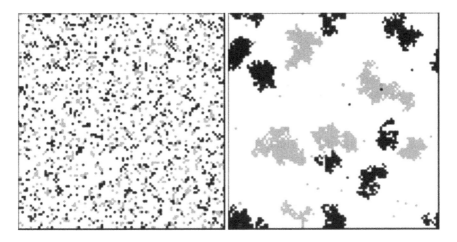

**Fig. 2.1** A demonstration of stigmergy—showing how ants sort items in a colony. The *shaded* indicates different kinds of objects within the colony. At *left* is an unsorted nest; at *right* is the nest after ants have been active for some time

wide variety of cases where simple behaviour of many organisms combines to produce an intelligent result.[6]

Although we like to think of ourselves as intelligent individuals, we are also creatures of habit. Just as ants use simple rules to guide their behaviour, we use simple rules to help us to get through our daily lives. Rules of thumb help us cope with life's complexities. Here are some famous examples:

- *A stitch in time saves nine.* (Traditional proverb)
- *Never look a gift horse in the mouth.* (Old saying)
- *Do unto others, as you would have them do unto you.* (Christ's golden rule)
- *Buckle up and live.* (Road safety slogan)
- *Neither borrower nor lender be.* (From *The Merchant of Venice* by William Shakespeare)
- *Be prepared.* (Motto of the Boy Scouts Association)

According to the French psychologist Jean Piaget, children learn to respond to the world by adopting *schemas*.[7] In effect, these schemas are like recipes: recipes for interpreting things and recipes for doing things.

As an example, think of a situation that you often encounter, say entering a room. You may not be aware of it, but you probably have a schema for entering a room. When you walk into a room, what do you expect to see? "Well of course," you would say, "there will be four walls, a floor and a ceiling. There would also be at least one door, and usually some windows." This is your basic mental model of what a room should look like. There are many possible refinements, such as lights and furniture, and details that depend on what sort of room it is. In a kitchen, for instance, you would expect to see a stove, cupboards and a sink. In a bedroom you would expect to find a bed.

So when you enter a room you mentally tick off the features you see against your mental model. Once you have ticked off an unimportant detail, you can forget it. What really stands out is a room that does not fit your model. Walk into a room with a wall or ceiling missing and you are sure to remember it. A kitchen with a bed in the middle is sure to seem unusual.

Children learn by building on and adapting the schemas they have already acquired. The main processes involved are known as *accommodation* and *assimilation*. Accommodation occurs when a child encounters a novel situation and finds a new pattern of behaviour, a new "schema", to deal with it. Going to a party, for instance, would not only involve a schema for entering a room but also schemas for meeting new people and various other interpersonal situations. Assimilation occurs when a person encounters a variation on a known situation and adapts an existing schema to deal with it. As a simple example, take the case of a schema for entering a room. By assimilation, this schema can be adapted and generalized to deal with entering all manner of rooms, including bedrooms, bathrooms, kitchens, auditoriums, and even cupboards. By accommodation, it can also be incorporated into schemas for entering a house, entering a theatre, coming home, visiting friends, and so on. In this way children acquire an increasingly rich repertoire of schemas for interpreting and responding to situations they encounter. Piaget proposed that another process, known as *equilibration*, ensures a balance between accommodation and assimilation.

Just as simple rules of behaviour lead to the organisation of ant colonies, so too the individual behaviour of people leads to the self-organization of human societies. Most of the rules that we follow are not at all obvious. For the most part, we are not even conscious that we are following them. And yet we see the results all around us. We see them in class distinctions, in real estate booms, and growing demand for tertiary education. We see them in Internet bubbles, stock markets crashes, in food fads, in football clubs, in discos and in a hundred and one other aspects that we take for granted. Despite all our laws and law enforcement, the orderly running of society would be impossible if people did not adhere to certain basic principles. Anarchy would quickly take hold if most people were not conditioned to be honest, to work hard, to be polite, and to keep their surroundings clean and tidy.

To make it clearer that we live our lives according to simple rules, consider the following question. What happens when you get up and go to work each morning? Most people would answer this question with a list something like the following:

> *get out of bed*
> *have a shower*
> *get dressed*
> *have breakfast*
> *pack a briefcase*
> *leave home*
> *catch the bus*

Of course, the details will vary from person to person, and even from day to day. But we can all identify with this sequence of events. Most of us have a routine that we follow on a regular basis. The point is that we are all creatures of habit. The above sequence is essentially a program for getting up and going to work. What is more, each step in our "going-to-work program" has substeps. Take just one step, say having breakfast. A typical "subroutine" for having breakfast might be as follows:

> *get out cereal, milk and cutlery*
> *fill bowl with cereal and milk*
> *heat bowl in microwave*
> *sit and eat cereal*
> *clear breakfast table*

Each of these steps can again be broken down into a yet finer repertoire of habitual routines that we use. Eating cereal, for instance, might follow a sequence something like this:

> *dip spoon in bowl*
> *remove spoonful of cereal and milk*
> *put spoon in mouth*
> *chew and swallow cereal*
> *if bowl contains more cereal*
> > *then repeat the above steps*
> > *else finish eating*

If we know exactly what is going to happen, as we do when eating breakfast, then we can work out a "program" that tells us exactly what to do at every stage. However, most real world situations are subject to many events and influences, most of which are unknown to us. Rules of thumb are recipes for coping with this uncertainty.

To take an example, suppose you are visiting a city that you have never been to before. You want to reach the city centre. If you have a road map, then you can plan exactly how to get there, which way to turn at every corner. But if you have no map, what do you do? The first time I visited Edinburgh, I had driven there by car from London. I had no map, but knew that my hotel was on the north of the city centre. So I used some simple rules of thumb: follow roads that lead more or less towards the centre but avoid minor streets and roads. This procedure worked so well I found the place without even resorting to asking locals for directions.

Another example is trying to reach the top of a mountain in heavy fog. You cannot see the top, so you just keep heading uphill. Of course, rules such as this are not fool proof. By going blindly uphill you could wind up on a knoll, not the peak, or reach a saddle where the ground is level in every direction, so it is no longer clear in which direction you should head. Nevertheless the rules are useful because they work in most cases.

We see similar issues in human situations. A rule like "always be polite to strangers" is a recipe for minimizing the risk of unintended conflict when people first meet. A proverb like "A stitch in time saves nine" is a recipe for preventing minor problems from growing into big ones.

The problem with simple rules is that they are based on assumptions, and assumptions often fail. If any of the items we need in our morning routine are missing, then the routine fails and we need to adapt or fall back on some alternative. Most of the time, having to adapt is mere inconvenience, but applying learned or habitual responses in inappropriate circumstances can have all kinds of unfortunate results. We shall see many examples in later chapters.

Another result of habitual behaviours is that they may have unintended side effects. Rules make life simple. But in following them, people ignore life's complexities. As we saw earlier when we enter a room we mentally tick off details that are unimportant. We then ignore them. But sometimes the details we ignore are important. By ignoring things that do not immediately concern us we ignore the effects that those things might have.

Unexpected trends emerge because we ignore the side effects that our actions may have. That extra salt on your food might kill you one day. Those long showers you enjoy might contribute to a water shortage. Potential consequences of our actions are often unseen and unheeded, until they jump up and take us by surprise. In the following sections, we look at just a few of the ways in which unseen side-effects can lead to surprising outcomes.

## The Human Ant Heap

The success of left-handers in sport, and spiralling real estate prices are just two simple examples of the many, many ways in which social trends emerge as unconscious by-products of human activity. In this book we will see why it is that, like the ants, we create unexpected social patterns and problems. The chapters that follow fall roughly into three parts. We begin in Chaps. 3–6 by looking first at people as individuals then at the ways we interact with others, and with the world around us. This sets the scene for understanding why side effects are unexpected. In Chaps. 7–13, we then look at society as a complex network of relationships and interactions, and some of the many ways in which complex networks lead to unpredictable outcomes. Having explored the human and network sources of unexpected results, in Chaps. 14–20 we conclude by looking at how they influence some major social issues.

## End Notes

[1]From the poem *To a Mouse* (1786) by Robert Burns.
[2]The way ants go about sorting their nests provides motivation for the Ant Sort Algorithm, which is discussed in Chap. 12. The Virtual Laboratory has a demonstration online at www.vlab.infotech.monash.edu.au.

[3]Feedback is a process in which the output of a process becomes new input back into it (see notes for previous chapter). See Chap. 10 for a discussion.

[4]Wilson (1975), p. 186. For a demonstration of how stigmergy works, see the Virtual Laboratory ant sort, at www.vlab.infotech.monash.edu.au.

[5]See for example, Bonabeau et al. (1999).

[6]One of the first studies to demonstrate the emergence of spontaneous order from simple behaviour was work by Hogeweg and Hesper (1983). They showed that the organization of bumble bee colonies emerged as a result of simple behaviour by many individual bees interacting with their environment.

[7]Piaget (1972), Block (1982).

# A Tangled Web

**3**

*The whole is greater than the sum of its parts.*[1]

**Abstract**

People like simple solutions to life's challenges, and reject complex solutions. However many things in life are complex. By seeking simple solutions we overlook important connections. Interactions and relationships between people and events form networks of connections. It is the richness and variety of those connections that creates complexity. Looking at networks of connections helps us to understand complexity and to see how unexpected situations can emerge from patterns of connections. Practices such as standards serve to decrease complexity in industry and large organizations. A universal winning strategy in competition is to restrict an opponent's network of possible actions; good luck is a by-product of expanding one's network of options.

## Keep it Simple

In April 2010, military planners gave a briefing for General Stanley McChrystal, commander of NATO forces in Afghanistan.[2] In the course of explaining American military strategy, they presented a slide showing all the groups involved and their relationships to one another. However, the diagram contained such a complex tangle of lines that it looked like spaghetti.

"When we understand that slide," said the general, "we'll have won the war."

The slide, and the general's remark, spread like wildfire over the Internet. In many quarters it was taken as a symbol of the failure of American policies and planning in Afghanistan. Amongst the public at large, it was taken for granted that a real solution would be simple.

D. G. Green, *Of Ants and Men*, DOI: 10.1007/978-3-642-55230-4_3,
© Springer-Verlag Berlin Heidelberg 2014

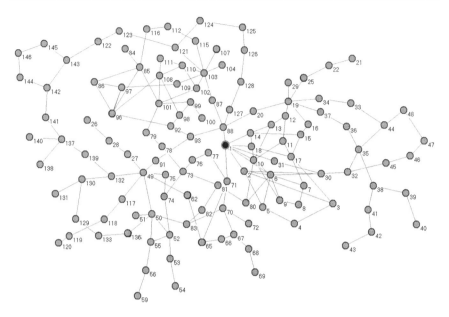

**Fig. 3.1** A complex system. The network of connections in the limits to growth model of world population. The *black dot* at the *centre* represents world population. *Other dots* represent socio-economic and environmental factors. The *lines* indicate causal relationships. For a detailed view of a part of the model, see Fig. 10.2

People want things to be simple. The assumption that things are simple, that real solutions to problems must be simple, is a universal theme in the way people deal with the world around them. They reject complex solutions. Even a suggestion that things really are complex is abhorrent. In political life it is tantamount to suicide.

In 2001, the Australian Labor Party, which was then in opposition, unveiled a plan to turn the country into a "Knowledge Nation."[3] However, the Government ridiculed the plan, focussing their criticism on a diagram in the proposal that showed a complex pattern of relationships between the many organizations involved.[3] The tangle of lines in the diagram led to the plan being dubbed "Noodle Nation". This joke contributed significantly to the Party's heavy defeat in the federal election that followed.

Even amongst scientists there is a tendency to reject complex explanations. In 1968, a group of scientists embarked on a project to understand the long-term consequences of interactions between population growth, socio-economics and the natural environment. Calling themselves the *Club of Rome*, they put together a set of simulation models of world dynamics, which came to be known as the *Limits to Growth*.[4] These models aimed to trace all the interactions between world population, economics, education, agriculture, and the environment (Fig. 3.1). Projecting current trends forward, they came up with a number of disturbing predictions. These included alarming explosions and crashes in world population, famines, runaway pollution and other disasters before the end of the twenty-first century.[5]

Needless to say, such dire predictions provoked both alarm and controversy. The models were subject to intense scrutiny.[6] The Limits to Growth model annoyed readers in two ways. First, its predictions of doom were alarming. If we are headed for social and environmental disaster within a couple of generations, then we need to start taking action now to avoid it. Secondly, the model was extremely complex. People were used to simple models in which it is easy to see that "A causes B". However, in a model as complex as Limits to Growth, it was difficult to see why particular kinds of behaviour emerged.

Rather than face the unpleasant truths embodied in the model, many people sought ways to reject its message. They attacked the model, picking at details to discredit the whole. Nevertheless, the study did highlight the danger of complacency about the future. It also helped to turn environmental issues into a matter for public and political debate. Several decades later, some of its scenarios have proved to be fairly accurate so far.

There is a well-known basis for people's difficulty in coping with anything complex. In 1956, the psychologist George Miller found that short-term memory imposes limits on our ability to process information.[7] Suppose that you have just been shown the correct responses to a number of different signals. It might to type a particular button for each colour that flashes on a screen. Then it turns out that most people are able to response correctly when there are four to eight alternatives, but make increasing number of mistakes when there are more choices. Remember that this applies to short-term memory. This does not mean that we cannot eventually learn more choices, but long-term memory needs to be involved.

This research became known as *Miller's Law*, which states that short-term memory can hold only $7 \pm 2$ chunks at one time. Here a chunk is any item with meaning that can be recognized. So chunks could be individual letters, but also entire words, because they have known meaning. Miller's law implies that confronted with even a moderately complex network, our brains could not hold all the details in short-term memory. However, if the network includes recognizable chunks, such as branches, then larger networks can be recalled.

## A Tangled Web

The world is big. The land surface of the Earth exceeds 130 million $km^2$.[8] At the time of writing, the 192 member countries of the United Nations have a total population of more than 7 billion people, and growing fast.[9]

We meet similar issues of size in dealing with the living world. The human brain contains about 40 billion neurons ($4 \times 10^{10}$ or 40,000,000,000). The human genome contains about 3.3 billion base pairs ($3.3 \times 10^9$ or 3,300,000,000).[10] And the human body contains perhaps as many as 100 trillion cells (i.e. $\sim 1 \times 10^{14}$ or 100,000,000,000,000).

The Earth is home to a large but uncertain number of species. By 2004 the world's biodiversity databases recorded 1.75 million species that had been described scientifically.[11] However, literally millions of species have not yet been recorded.

Anyone confronted with more than a tiny fraction of the world can easily be daunted by the sheer volume of detail. Computers have gone a long way towards helping us to cope with such overwhelming detail. In the 1950s, IBM developed a pioneering computer called STRETCH that was capable of 5,000 floating point operations per second. In 2008, IBM's Roadrunner supercomputer became the world's first computer to achieve a processing speed of 1 Petaflop (1,000,000,000,000,000 floating point operations per second).[12] This represents an increase in processing speed by a factor of over 200 billion in a little over 50 years. At the time of writing, a typical home computer has 8 GB of memory, up to a TB of data storage and is capable of processing speeds that exceed 10 GFLOPS (10,000,000,000 floating point operations per second).

As the above figures show, the memory capacity of modern computers approaches and even exceeds some of the huge systems described above. A database that provided (say) 1 KB of data about every person on Earth would require just over 6 TB of storage. This is well within the capacity of any major computing facility.

Storing data about a system is one thing. Actually modelling that system in the computer is another. To capture the real system, we need to capture it in all its complexity. This means, for example, not just storing data about all the people in the world, but also about their relationships and interactions with one another. To appreciate this, we need to look briefly at the nature of complexity.

Sheer numbers do not mean complexity. What makes true complexity is not size, but the number of ways in which different parts of the system can combine and interact.

Have you ever been at a small social gathering and it turned out that two of the people there shared the same birthday? This is a well-known party game. There are 365 days in a year, so in a small group it would seem unlikely that anyone would just happen to have their birthday on exactly the same day as someone else. Well, that is certainly true for any particular pair of people. But at a party there are many different pairs. If just 23 people attend the party, then there are 253 different pairs, and it turns out that there is a 50:50 chance that at least one pair of guests have birthdays on the same day.

We can begin to understand this by looking at *combinatorial complexity*—the way in which the number of possible combinations grows as the number of items increases. Suppose that from a pool of 10 people, you need to select one person to do a job. Then of course, there are 10 possible choices. However, if you need to choose a pair of people, then there are 45 possible choices.[13] If we label the people A, B, C, D, E, F, G, H, I, J, we can list all 45 combinations as shown in Table 3.1.

Now 45 alternatives may not be that many, but if we increase either the pool of people or the size of the group we want to select, then the number of combinations grows very rapidly (Table 3.2). To select 5 people from a pool of 100 candidates,

**Table 3.1** All 45 combinations of pairs from the pool A–J

| A B | A C | A D | A E | A F | A G | A H | A I | A J |
|-----|-----|-----|-----|-----|-----|-----|-----|-----|
|     | B C | B D | B E | B F | B G | B H | B I | B J |
|     |     | C D | C E | C F | C G | C H | C I | C J |
|     |     |     | D E | D F | D G | D H | D I | D J |
|     |     |     |     | E F | E G | E H | E I | E J |
|     |     |     |     |     | F G | F H | F I | F J |
|     |     |     |     |     |     | G H | G I | G J |
|     |     |     |     |     |     |     | H I | H J |
|     |     |     |     |     |     |     |     | I J |

**Table 3.2** Growth in the number of possible combinations

Size of group selected

| Pool size | 2 | 5 | 10 | 50 |
|-----------|-----|-----|-----|-----|
| 10 | 45 | 252 | 1 | – |
| 100 | 4,950 | 75,287,520 | $1.73 \times 10^{13}$ | $1.01 \times 10^{29}$ |
| 1,000 | 499,500 | $8.25 \times 10^{12}$ | $2.63 \times 10^{23}$ | $\sim 2 \times 10^{85}$ |

there are over 75 million possible combinations. To select 50 from a pool of 1,000, the number of combinations is truly astronomical: 2 followed by 85 zeroes.

As the table shows, the number of combinations we can select from a collection of things can be astronomical. But if we take order into account, then the number of possible permutations grows at an even more alarming rate. One of the best known examples concerns a travelling salesman who needs to visit a number of towns.[14] Knowing that time is money, he wants to find the shortest route, and avoids visiting any town twice. For 10 towns, there are already 3,628,800 possible routes. It turns out that if there are just 29 towns, then there are so many possible routes our salesman can take ($8.8 \times 10^{30}$) that even using all the computers in the world it would take longer than the age of the universe to check them all.

However, even the rate at which the number of possible orderings increases pales when we look at patterns of connections between objects. A network consists of objects (also called "nodes" or "vertices") with links (also called "edges" or "connections") between pairs of objects. You could think of it as a "ball and stick" model. A road map, for instance, is a network that shows towns (the objects) linked by roads (the edges). A family tree is a network in which the people form the nodes and parent-child relationships form the edges.[15]

Even with small numbers of objects, the number of possible networks that can be formed grows extremely rapidly. To begin, suppose we have just three objects (Fig. 3.2a). Call them A, B, and C. Then from these we can form 8 possible networks: 1 network without any edges, 3 with just one edge (A–B), (A–C), (B–C),

**Fig. 3.2** Complexity of networks. **a** All simple networks of three nodes. **b** Growth in the number of possible networks as the number of nodes increases. The *vertical axis* is on a logarithmic scale (e.g. 20 means the number 1 followed by 20 zeroes). As the *graph* shows, over 1,027 different networks can be made from just 10 nodes

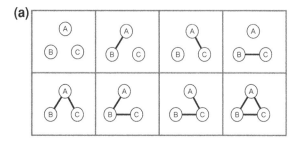

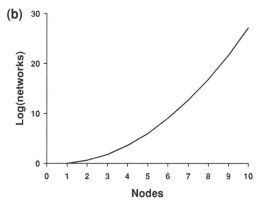

3 with two edges (A–B, A–C), (A–B, B–C), (A–C, B–C) and 1 network with 3 edges (A–B, A–C, B–C).[16] For 10 objects, the number of possible networks rises to more than 3.5 trillion (Fig. 3.2b).[17] Imagine that those networks represent relationships among groups of people, and you will see how complex social relationships can be!

Networks are important because they underlie many of the complexities we encounter in modern society. Besides our personal networks of friends, family and colleagues, there are networks in every kind of infrastructure. They include the Internet, transport networks, socio-economic networks, telecommunication networks and supply networks, to mention just a few. But these are physical networks. The importance of networks extends far deeper than mere physical connections. Networks also turn up in the ways we organize ourselves and in the patterns of events that unfold over time. We will come back to look at social networks more closely in Chaps. 13 and 14. In the remainder of this chapter we shall look at some of these ways networks occur in our lives.

## A Flock of Birds

New Year's Eve of the year 2000 ended with a scary experience for my young family. A vast crowd gathered on the bank of the Yarra River in my home town Melbourne, to count down to the new millennium. After the fireworks ended everyone streamed off the riverbank heading into the city centre to make their way home. Literally hundreds of thousands of people moved to cross over Princes Bridge at the same time. Within minutes the bridge was clogged. Along with my family, I became trapped in a slowly moving sea of humanity. People were packed so tightly there was no freedom of movement at all. We were jammed tightly like sardines, crushed against complete strangers on all sides. It was impossible to walk in any normal sense, so we had to waddle, inching forward in time with the rest of the crowd. I knew that if any of us so much as stumbled, we would be trampled to death in seconds. Fortunately, there were no accidents that night and we emerged unscathed.

Crowds are networks that form when large numbers of people interact with those around them. They occur in all sorts of situations, from crowds of spectators cheering at the final of a major sports event to commuters streaming like a river through Tokyo Railway Station.

Many other kinds of networks also arise spontaneously and for a short time only. A flock of birds, for instance, appears when flying birds coordinate their behaviour with one another. Each bird interacts with its neighbours, pointing in the same direction, but keeping a safe distance apart. In this way they form a network that lasts as long as their migration.

Other networks are more long-lasting. A family tree, for instance, is a permanent set of connections between people, whether they interact with one another or not. A network of friends consists of people who see each other more or less frequently. For any particular group of people, the exact network changes over time as individuals make new friends and lose contact with others.

## A Game of Chess

A more subtle, but nevertheless important, kind of network appears as events unfold. As an example, take the problem of travelling through the countryside from town A to town B. The roads (or railway lines, or canals) form a network of connections between towns. The route that we take from A to B can be seen as a map, not just of the countryside, but of our change in position over time.

Society is full of dynamic networks. Some are crucial to the smooth running of our world. One kind of dynamic network is the supply network by which goods are produced and delivered. In a computer, for instance, there are many different components. The manufacturer obtains many of these components (e.g. the central processor) from other manufacturers who produce specialized parts. These manufacturers in turn obtain materials (metals, plastics) from companies that produce them. And so it goes on. For any product that we use, there is a supply network of companies that contribute towards putting that product together.

In the same way, events in our lives form networks of experiences. Our decisions at any time determine which route we shall take through life. As a smaller example, consider a game of chess. At any given time, the arrangement of pieces on the chessboard defines the state of the game. At the start, all the pieces are lined up on opposite sides of the board. When a player moves, that move transforms the board from one arrangement to another. So we can think of a game of chess as a pathway through a dynamic network consisting of all the possible arrangements of pieces on the chess board.

If we view chess as a network of possible states of the chessboard, then a game of chess is a pathway through that network. Decisions made by the players determine the course the game takes. In the same way, the decisions that people take determine the path their lives take through a vast network of potential futures. The decisions made in boardrooms determine the path that a company's fortunes take through a vast network of financial possibilities.

## Batteries Not Included

When our distant ancestors started using tools, society began an upward spiral of increasing complexity. Tearing meat off a carcass is difficult, even with a flint knife. So began a process of refinement that produced better and better flint blades. The use of blades, for instance led first to sharper blades, then to blades adapted for different purposes, and to tools to help make blades. The increasing complexity and skill required to make blades turned flint knapping into a specialized craft. Trading began and the proliferation of tools, techniques and products led to the first supply networks. Modern society has taken this approach and extended it on a vast scale.

The problem with supply networks is that as they become larger, they can also become increasingly fragile. Take the case of a supply chain, for instance. Suppose that company A supplies something to B, B then supplies C, C supplies D and so on until eventually Y supplies some essential item to Z. A chain is only as strong as its weakest link, so if one supplier in the chain fails, then the entire chain fails. The problem is solved, at least to some extent, by having a network in which there are several potential suppliers at each point in the chain.

Problems of this kind highlight the importance of understanding how properties of complex networks affect the stability of supply, as well as many other aspects of society. An immediate question is how to make supply networks more robust? Industry has found one answer: reduce the complexity of the supply network. One way to do this is standardization.

Take batteries for instance. There are well-defined standards  for common batteries in home use (AA, AAA etc.). So if you have a device that uses a standard battery, then obtaining replacement batteries is easy. This simplicity also helps manufacturers. If their product uses a specialized battery, then the company must devote time and effort to manufacturing the batteries as well. Alternatively (and often more likely) they could outsource the job to a specialist company and purchase the batteries from them. However that means supply of their product depends on a sole supplier. If anything happens to that supplier, then their product is in trouble. Likewise, the company that produces a specialized battery becomes dependent on a single client.

If instead of a specialized battery, the company uses a standard battery in its product, then the matter becomes far simpler.[18] The manufacturer will most likely have several suppliers to choose from. If one supplier fails, there are others to fall back on. Likewise, the supplier is not dependent on a single client to purchase the batteries. Other companies will want to use them too. Battery technology is evolving fast, with a plethora of different methods being used to generate and store power. The presence of standards helps to insulate those changes for the specialist industries that need them.

We can see similar examples in information technology. Take image formats for example. Anyone who uses a digital camera is familiar with some formats, such as JPEG. But computer programmers have developed literally hundreds of other ways of encoding images. Often there is a need to convert an image from one format to another. If there are (say) 100 different formats, then there are nearly 10,000 (=100 × 100) possible transformations of one format into another. Writing software to handle so many possible conversions would be an enormous waste of time and effort. Fortunately it is not necessary. All that is needed is a standard interchange format (call it $X$). For each other format you write a program to convert to the standard and another program to convert from the standard. So to transform any format $A$ into any other format $B$, first you convert $A$ into $X$, then $X$ into $B$. This way you need only 200 conversion programs, not 10,000.

In 2009, a Japanese study looked at the robustness of two major industries: automobiles and computer software.[19] Given the advanced nature of computer software and hardware, the researchers expected that the software industry was more developed. However the study found the exact opposite to be true. Having existed for much longer, the automobile industry had long ago weeded out inefficient players and evolved robust standards and supply networks. In comparison, the software industry was still evolving, with many small players and a relative lack of standards.

Some standards are so familiar that we do not even think of them as standards. One of the earliest, and certainly the most widely used standard in every society is money. Money seems to have appeared independently in every society to satisfy the need for some universal token of value. It appeared wherever trading networks grew too large for peer to peer consensus over bartering.

Like other interchange standards, money simplifies the problem of comparing items for exchange. In bartering, you need to agree on the relative value of each item against each other item. One stone axe may be worth ten rabbit pelts, for example. The problem is that this becomes extremely complex, and contentious, very quickly. If 100 different kinds of items were offered in a large market place, then a bartering system would require about five thousand separate rates of exchange. This would be a recipe for endless haggling and animosity. On the other hand, if everyone expresses the value of their goods in terms of a common currency, then working out the rate of exchange is much simpler. You can easily see you need to sell 10 rabbit pelts at 10 shekels apiece to afford an axe priced at 100 shekels.

The importance of simplifying values in this way was driven home to me some years ago on a visit to Brazil. The country at that time was undergoing runaway inflation. A note for 100,000 cruzeiros might drop to half its buying power in a matter of days. To cope, many shopkeepers preferred to deal in US dollars and other foreign currencies. Another, more basic problem was how to put price labels on goods. Changing thousands of price tags on individual items every day was impossible. They solved the problem by labelling goods in price categories A, B, C etc. That way they needed only a single notice on the counter listing the current price for each category.

## Winning Strategies

You see it often in tennis matches. One player gets control of a point by hitting the ball deep to the right side of the court. Her opponent races to the right and just manages to scoop the ball back over the net. The first player calmly thumps the ball back again, this time to the left side of the court. Her opponent races after it, swings desperately and, just clipping the ball, sends it flying high into the air and it lands amidst the spectators.

This is an example of a universal winning strategy in sport. You push your opponent beyond the point where she can cope, either physically or mentally. This is a special case of a winning strategy that applies to many games. To understand it, think of a game as a sequence of moves by the players. If your opponent has a choice of many moves, then there are many different paths that the game can follow. On the other hand, the fewer moves a player has at any time, the more predictable they are and the easier to control. So the most general winning strategy is to restrict your opponent's moves while simultaneously maximizing your own choice of moves. Perhaps the best known example occurs in chess, where a player forces an opponent to make certain moves by repeatedly putting the king in check.

Thinking of activities as networks of moves helps us to understand other kinds of strategies. In some activities, the network of possible moves changes as the "game" progresses. This often makes it necessary to adapt your strategy to cope with these changes.

As a simple example, think of how you solve a jigsaw puzzle. Essentially the puzzle consists of finding pieces that fit together. However, the way you do that changes as the puzzle progresses. A typically large puzzle might contain 1,000 pieces, arranged in 25 rows of 40 pieces each. So if you tried to match up pieces purely at random, then there are 499,500 possible pairs of tiles to compare. But each tile has 4 sides, so there are 7,992,000 possible pairs of edges to compare and only 1,935 of them are correct. To check every combination individually, at one per second would take nearly three months, without any rest or sleep at all. But this is not the way you tackle a jigsaw. Almost everyone starts by picking out the edge pieces. There are only 126 of them. Then you group pieces that have distinctive colours or patterns into separate piles. Within each of these groups, joining up pieces is like a very small jigsaw, a much easier problem.[20] As you form more and more groups of tiles, you can start joining the groups together. When the overall jigsaw starts taking shape, the problem becomes one of filling in spaces, which get smaller and smaller.

The nineteenth century saw gold rushes in many parts of the world. Millions of years of erosion and other geomorphic processes had left gold scattered widely across the landscape. The pieces ranged from dust to huge nuggets. In the early days of each gold rush, this distribution of gold pieces lent itself briefly to thousands of prospectors being successful. But as the gold ran out, there were no more easy pickings and only companies with resources to dig mines along seams could make a profit.

Ants do the same thing when exploiting resources in their environment. The first task is to locate items of food, which is scattered widely across the landscape. The workers become prospectors foraging in different directions, gleaning tiny scraps of food here and there. If an ant comes across a large resource, then the ants reorganize themselves dramatically. Instead of searching randomly, they follow pheromone trails and form mini highways along which ants pass back and forth as they travel between their nest and food they have discovered.[21]

In the same way, businesses have to change the way they operate as they grow larger. In a small company, a single person can often handle every aspect of the business. This may be the best way of ensuring coordination and efficiency. But in a large company, no one person can cope with the complexity. There are just too many details interacting with each other. The company needs to reorganize to survive. It divides its work into specialized functions (manufacturing, sales etc.), each managed by different people. It also needs to implement standards and procedures to coordinate the operations of all the different sections.[22]

## On Being Lucky

The first time I visited London, I thought that I knew no one at all in the city. But walking down the Strand for the very first time, I bumped into one of my Canadian friends. Unknown to me, he had moved to London to study a few months earlier. The chances are small that any particular person you know will be visiting the same place at the same time as you do. But if you have a wide network of friends and acquaintances, then the chances are much greater that some friend (or a friend of a friend) will be visiting some place you visit at the same time. If you visit places where friends are thick on the ground, then it can become almost inevitable that you will bump into someone you know. This can lead to unplanned traditions, such as school kids converging on a shopping mall after school.

The above story provides an important clue about how to be lucky. The richer a network of connections is, the greater the chance of an event happening by luck. So to be lucky, simply increase the number of connections that lead to your goal. If you want to find a girlfriend or boyfriend, then maximize the opportunities to meet potential partners. Go to places and events where you will see other people. Get out to parties. Make lots of friends who can introduce you to people. If you want to win prizes, then enter lots of contests. Your chances of winning any one contest may be small, but if you enter enough, then the chances of winning something, somewhere are good.

If some people seem to be lucky, perhaps they are. Lucky people tend to make their own luck. Whether by design, or simply by habit, they maximize their chances of succeeding at whatever they may be doing. The way they do this is not only by creating opportunities. It is also by recognizing opportunities. The psychologist Richard Wiseman has argued that people who are considered lucky are often more observant than people who are unlucky.[23] Lucky people notice more about their environment. They see opportunities and act on them. Unlucky people are often less observant.

Put in reverse, it follows that observant people are less accident prone. They are more likely to notice danger than people who are non-observant. And of course, any particular person will be more or less accident prone at different times and in different conditions. If you are tired, intoxicated, or distracted, then you notice less and become more accident prone.

## End Notes

[1]Anonymous saying.

[2]The incident was described by Elisabeth Bumiller in an article titled "We have met the enemy and he is PowerPoint" published in *The New York Times* (World) April 26, 2010. http://www.nytimes.com/2010/04/27/world/27powerpoint.html.

[3]Contributing authors (2011). Knowledge Nation. *Wikipedia*. http://en.wikipedia. org/wiki/Knowledge_Nation.

[4]Forrester (1971).

[5]Meadows et al. (1972).

[6]Cole et al. (1973).

[7]Miller (1956).

[8]FAO (2011).

[9]United Nations, Population Division (2012). *World Population Prospects, the 2012 Revision*. www.esa.un.org/unpd/wpp/. Downloaded 15 July 2013

[10]Human Genome Project. http://www.ornl.gov/hgmis/.

[11]Canhos et al. (2004). Global biodiversity informatics: setting the scene for a "new world" of ecological modelling. *Biodiversity Informatics* 1, 1–13. https:// journals.ku.edu/index.php/jbi/article/viewFile/3/1. Downloaded 21 October 2013

[12]Top 500 Supercomputer sites (2010). http://www.top500.org/.

[13]In general the total number $M$ of combinations when $R$ objects are drawn from a pool of size $N$ (without replacement) is given by the formula

$$M = {}^{N}C_R = \frac{N!}{R!(N - R)!}$$

Here $^{N}C_R$ is a shorthand way of referring to the number of different combinations of $R$ objects that can be drawn from a pool of $N$ objects. And

$$N! = N \times (N-1) \times (N-2) \times \cdots \times 2 \times 1.$$

[14]The Travelling Salesman Problem (TSP) is a classic "hard" problem in optimization. The figures mentioned here refer to a "brute force" approach in which every possible case is checked. Many techniques have been discovered to speed up the search. In practice, real salesmen are likely to simplify the problem by applying practical considerations, such as basing themselves in a number of centrally located towns and making a series of short tours to several neighbouring towns.

[15]Of course a real family tree also includes marriage as another type of link between people.

[16]Here we are ignoring the direction of edges. That is we take A–B and B–A to mean the same thing. This would not be the case in a family tree. For instance, if A–B means "A is the mother of B", then its meaning would be very different from B–A.

[17]For $n$ nodes, the number of networks $G$ is given by the formula $G = 2^{n(n-1)/2}$.

[18]In practice it may still be complicated. If the standard is a proprietary one, then any manufacturer may need to purchase rights to use it from the owner.

[19]Hiroyasu Inoue (Osaka Sangyo University, Japan), Analyses of procurement networks in automobile and software industries. *Proceedings of Complex'09*, 2009/11/5.

[20]This is an example of "divide and rule" which we will look at more closely in Chap. 8.

[21]See for example, Bonabeau et al. (1999).

[22]We look at this "divide-and-rule" approach more closely in Chap. 8.

[23]Wiseman (2003).

# The Eye of the Beholder

4

*One thing a person cannot do, no matter how rigorous his analysis or heroic his imagination, is to draw up a list of things that would never occur to him.*[1]

**Abstract**

The complexity of the world around us means that unwelcome events are often sudden and unexpected. Limitations of our thinking contribute: our models of the world around us are necessarily limited in scope and detail. Also human perception is local, so we are not able to see everything that might affect us. Being unable to understand the complexity that leads to misfortune, people look for simple explanations. One of these has been to attribute misfortune to the will of the gods. Another is to look for human scapegoats, which in extremes has led to witch hunts. Confirmation bias—ignoring contradictory evidence—distorts people's perception of events and reinforces prejudices in these situations.

## The Glass Half Empty

*1:15 p.m., Friday, Hometown USA.* It is lunchtime and during his break George visits the bank to pay some bills. The bank is crowded and George finds himself at the end of a long queue. He glances anxiously at his watch. He has to be back at work by 1:30. Peering over the heads of the people in front of him, he sees that way up the front of the queue a little old lady is slowly counting out coins to deposit. Looking around, he sees that the next queue is moving much faster. The guy in front of him moves over to the shorter queue and George follows. The queue continues to move. With a smug sense of relief, George notices that he is several places ahead of where he had been in the other queue. Then the little old lady who was holding things up leaves. Meanwhile a young couple reach the front of his present queue. Moving to the counter, they pull out a stack of papers and start arguing with the teller. George's queue grinds to a halt, but his original queue is now moving rapidly.

George is still stuck behind three other people when he realizes that a lady leaving the teller was originally behind him in the first queue. It suddenly dawns on George that if he had stayed put, he would have been back at work on time.

"Why do I always get stuck in the slow queue?" he groans in frustration.

Poor old George. It seems that fate keeps dealing him a bad hand. Or does it? The truth is that George does not always get stuck in the slow queue at the bank. It only seems that way. As often as not, he winds up in the fast queue. But being in the fast queue is not frustrating, so he soon forgets about it. The only visits that do stick in his memory are the frustrating ones, the visits when he does wind up in a slow queue.

The first thing we have to ask when things seem to be bad, is whether they really are bad? Are things bad or is it just pessimism or paranoia? Often what seems to be bad or what seems to be getting worse, is simply a negative attitude or an over-reaction to what are really just random events. Attitude (e.g. bias, bigotry) often dominates over reason, leading to poor decisions.

Part of the problem is that most people, most of the time, do not have the advantage of complete knowledge of what is going on around them. They see only what is going on in their immediate surroundings. They never get the opportunity to see the big picture. Also, it is difficult, and often next to impossible to tease out all the complex chains of cause and effect that ripple through society. As George learned to his cost in the bank, it is often impossible to predict how events will play out in certain circumstances.

The problem is that when bad things happen, people need to rationalize those experiences. They tend to find simple cause and effect explanations that are based on their limited experience of reality.

George's little tale of frustration highlights the selective nature of perception. We remember things that stand out, such as long queues, because they do not fit easily into our scheme of how things should be. We forget all the times the queue moved quickly. And from these biased recollections of our limited experience, we formulate general notions about the world around us.

As we saw in Chap. 2, people respond to the world by adopting schemas: mental recipes for understanding and reacting in common situations. Schemas are crucial because they provide ready-made solutions to life's challenges. They help us to cope with the complexity of the world in two vital ways: they provide ready-made solutions to scenarios we often encounter and they allow us to ignore unimportant detail. Forgetting unimportant details is crucial. It helps our brains to cope with the flood of information we receive all the time. So after you see a ceiling you forget all the little details that are not important, such as its colour, or the location of lights. Of course, people's schemas will differ. An interior designer, for instance, would note much more detail about a room than most other people. What you do remember are details that are unusual or important. Walk into a room that has no wall and you are sure to remember it. This selective memory has important consequences.

The process of creating schemas contributes to habit formation. Once the brain forms a schema, repetition serves to strengthen it. Every room you enter calls up your "entering a room" schema, so we fall into the habit of applying this and related schemas as a matter of course. Such habits can lead to a phenomenon known as *confirmation bias*,[2] which contributes to many of the problems we will meet in later chapters.

Any belief we hold colours the way we interpret our experiences. If those experiences confirm that belief, then it strengthens that belief. If it does not, then it is ignored. Taken to extremes, our tendency to generalize from particular cases, combined with confirmation bias, can lead to prejudice. For example, a child is bitten by a dog. Generalizing, the child forms a belief that all dogs are dangerous. The child quickly forgets subsequent experiences with friendly dogs, but any encounter with an unfriendly dog recalls the experience of being bitten and serves to strengthen the prejudice. Likewise, being stuck in a slow queue while others are moving ahead confirms the belief that your queue will always be slow.

And just as it does with our impression of bank queues, confirmation bias sometimes leads to poor impressions in many other settings. Most of the time, the trains or buses we catch run on time, but sometimes they are late just when we are in a hurry. So we forget the hundreds of times they have been on time and develop the attitude that they are always late. In an office, I have seen cases of people who are hard-working, dedicated and doing an excellent job. But all that hard work gets ignored by the boss. Instead, one or two slip ups, as often as not caused by the boss himself, leads an angry boss to vent the opinion "She's hopeless. She always messes things up."

Schemas that we use to guide our interactions with the world around us are mental models. The use of models is so widespread and commonplace in society that we tend to take them for granted. And yet, shortcomings of models underlie many of the unexpected problems that confront us. So it is worthwhile looking at the nature of models a bit more closely.

## From Toy Ships to Fashion Parades

Throughout this book we shall need to refer many times to models. The word "model" encompasses a great many things and to understand the human ant hill, we need to have a clear understanding of what models entail.

In general, a model is any representation that preserves key features of the original. A toy ship, for instance, is a model of a real ship because it provides a fairly detailed idea of what a real ship looks like. In a fashion parade, catwalk models show what clothes will look like when you wear them. Well, actually, they are not meant to be entirely faithful: they try to flatter the buyer by showing them how they would like to look.

A road map is a model in which towns are shown as dots and roads are shown as lines joining the dots. Apart from names of town and roads, these maps usually include little else. Landscape features such as hills, houses, lakes and forests are largely ignored. For the navigator trying to find the correct turn off, they are unimportant.

Models in science and engineering take many forms. They can be miniature physical copies of the original, such as a model aeroplane or an architect's design. They can be diagrams, or formulae, or computer simulations. As we saw in Chap. 3, we can model complex systems using a ball-and-stick model. That is, as a network of nodes (balls) linked by edges (sticks).

Piaget's schemas are cases of mental models. As we saw above, they tell us what we expect to find in certain settings. They also tell us how to react. The idea of schemas is so powerful it has been adopted into computing as the basis for what is known as object-oriented modelling (or OO for short).[3] This treats systems as networks of inter-related objects. To take just one example, a book is an object; so is an author, and so is a publisher. Authors can write many books, but usually each book has a single author.

Object-oriented modelling borrows from Piaget's ideas of assimilation and accommodation. Assimilation is a process of generalization: adapting an existing schema to new situations makes it more general. Think about books again. The notion of a book is a more general class of object than (say) a novel, which is a special kind of book. Likewise, a book is simply a special case of a more general class of objects: publications, which also include magazines, maps and newspapers. Accommodation is a process of creating new schemas from existing ones. Object-oriented models do the same by creating new objects by combining existing objects. A book for example is an object that consists of smaller objects: a title page, contents, chapters and an index.

Models are normally used in one of three ways (which often overlap).[4] Models are used for *explanation* to help us understand what something is like, or how it behaves. They are often used in *prediction* to tell us what to expect in the future (or in different circumstances). Finally, models are used for *control* to tell us what a system should be like.

In this last case, it is not a matter of trying to make the model mirror the system, but making the system mirror the model. Governments tweak factors such as interest rates to try to make markets behave like their economic models. Road rules are aimed at making drivers behave according to a model of safe-flowing traffic.

One of the most important things to remember about models (whether formal or mental) is that they are limited. No model ever captures everything about a thing. Like the road map, models represent only those features that are essential to understand a system for a particular purpose. What they leave out is just as important as what they include. In later chapters we shall see that what models ignore can have widespread and sometimes devastating consequences.

Built into every model there are assumptions that affect whether or not they are valid and useful. These assumptions are often hidden, unintentional and subtle. To take a simple example, since the 1500s, the world's population has been growing at an accelerating rate. It reached 0.5 billion around 1565 and 1 billion around 1805.[5] By 1925 it had doubled to 2 billion and by 1975 it had doubled again to 4 billion. These figures show a pattern in which each doubling of the world's population took only half as long as the previous doubling. If we take this pattern literally and use it as a model to predict future growth, then by 2035 the world's population would be essentially infinite! This will not happen. For a start the model assumes implicitly that the doubling time can go on reducing until it is virtually nothing. This assumption ignores the time humans take to reach reproductive age. The numbers assume that babies reproduce immediately they are born! But even if we allow for (say) 25 years before the next generation breeds, the model would still predict a world population of around 120 billion by the year 2100. Again this will not happen because problems of food supply and waste would interfere long before then. In fact, population growth is already slowing in most western countries. By 2012 the world's population was not 10 billion, but approaching 7 billion. In other words, the model's assumptions are no longer valid.

In later chapters we will see several ways in which inadequate models lead to unexpected events and trends. In the following sections, we turn to ways in which people's attitudes lead them to adopt and retain false models. We will look at how ignorance, poor information and limited perception lead people to adopt poor models of the world around them.

## Seeing the Wood for the Trees

One of the highlights at any fair is the Ferris wheel. You sit in a seat attached to a large wheel. As the wheel rotates it takes you up into the air. The main attraction of the Ferris wheel is that it gives a different view of the world around you. As the wheel turns, your perspective changes. When you first take your seat you are surrounded by people, by noise and by all the paraphernalia of the fairground. Then as you rise, the world seems to fall away beneath you. The detail no longer dominates your vision. You see the entire layout of the fair ground. You even see how the fair ground sits within the surrounding landscape. As the wheel continues turning, you descend again. Once more you are surrounded by all the detail, but now your surroundings seem somehow altered. This is because you have had a glimpse of the larger context. The Ferris wheel allows you to see how your surroundings fit into the larger whole.

A lot of situations in life are like this. The complexity of detail can make it difficult to see the whole amidst the parts; to see the wood for the trees. As we saw previously, the desire for simplicity leads people to overlook or deny complexities. And when we cannot see the whole picture, only local details, we are susceptible to local influences. So people assume that what they can grasp with limited perception shows the way things are everywhere.

## In the Lap of the Gods

People want order. People need order. They want the world around them to be simple, not complex. They want certainty, peace and stability. They do not want confusion, strife, or change. They want to be able to control their lives. And therein lies one of the great conundrums of human existence. For the world around us is complex. It is constantly changing. It is often capricious, unpredictable and uncontrollable. The dichotomy between what we want and what we get colours much of our thinking. It influences much of our culture. It underlies many aspects of moral and religious beliefs. Anything that helps us to achieve our most cherished goals is seen as good. Anything that frustrates those ambitions is seen as bad.

Embedded in our beliefs, and even our language, are many age-old ideas about the nature of order and disorder. Roget's thesaurus links disorder with anarchy, chaos, derangement, and discord. It links complexity with agitation, trouble and turmoil. Storms, earthquakes, floods, famines, disease, and pestilence all disrupt the regular order of human affairs. Given the disastrous and unpredictable nature of natural events, people tend to associate order and disorder with natural forces.

Because they frustrate the desire for order, people also link complexity and disorder with evil. For instance, we say that "all hell broke loose" when things get out of control. Writers, such as Shakespeare, draw many associations of this kind:

> … But when the planets
>   In evil mixture to disorder wander,
>   What plagues and what portents, what mutiny,
>   What raging of the sea, shaking of earth,
>   Commotion in the winds! Frights, changes, horrors,
>   Divert and crack, rend and deracinate,
>   The unity and married calm of states
>   Quite from their fixture! …[6]

We also see similar allusions in songs, for example[7]:

> O Fortune, like the moon you are changeable, ever waxing and waning; hateful life first oppresses and then soothes as fancy takes it;
> poverty and power it melts them like ice.
> Fate—monstrous and empty, you whirling wheel, you are malevolent,
> well-being is in vain and always fades to nothing, shadowed and veiled you plague me too;
> now through the game I bring my bare back to your villainy.
> Fate is against me in health and virtue, driven on and weighted down, always enslaved.
> So at this hour without delay pluck the vibrating strings;
> since Fate strikes down the strong man, everybody weep with me!

The belief that evil forces lurk behind every calamity is so firmly entrenched that most religions personify chaos and disorder as divine spirits. The ancient Egyptians had Seth, the god of evil and chaos. In ancient Greek mythology, Eris, the goddess of discord rode a chariot behind Ares, the god of war, spreading confusion wherever she rode. For the ancient Romans, there was Discordia, goddess of dissent. Meanwhile, the Hindus had Vrta, god of chaos and Varuna, god of the sea and of order. And in Scandinavian mythology, Loki, the spirit of evil, killed Balder, their god of the summer sun. Christian beliefs have reduced things down to a simple duality. God represents order and goodness, whereas Satan, the devil, represents chaos and evil.

Humans are inherently egocentric. From childhood, all our experiences are centred on things going on around us. We also tend to anthropomorphize the world around us. People reveal this tendency in simple expressions such as:

My car knows where it's going.
  This machine doesn't like me.

Studies have found that people are more likely to anthropomorphize in situations where events are unpredictable.[8] And complex situations are likely to be unpredictable. When there is no obvious cause, unexpected events seem more like the doings of some capricious intelligence. We invent gods to explain events in nature beyond our control.

This tendency to ascribe deliberate purpose to nature is built deeply into human nature. There is some evidence that this tendency may have had its beginnings in our animal origins. In her observations of chimpanzees, Jane Goodall observed that during rain storms, adult male chimpanzees would carry out an elaborate "rain-dance" in which they swaggered about, broke off and waved branches at the clouds and hooted at the sky. It was as if they were defying nature in the same way they might confront rivals.[9]

Whenever people don't understand something, they try to ascribe a simple cause to explain it. Scientists wrestling with the problem of what makes things alive have often sought simple explanations. In her novel *Frankenstein*, Mary Shelley followed common thinking of the day by presenting electricity as the vital spark that turned inanimate matter into life. In like fashion, people have often tried to make sense of complexity in the world around them by assigning chaos and disorder to malevolent spirits.

The belief that chaos and calamity stem from divine intervention has flourished over the millennia because people did not understand the link between complexity and chaos in human affairs. In essence, attributing chaos to divine will is really a cop out. If misfortune is an "act of God", then no one is to blame. No one could have done anything to prevent it.

## Witch Hunts

During January and February 1692, in the town of Salem, Massachusetts, two young girls, Betty Parris and her cousin Abigail Williams, suffered a series of fits that alarmed their families and the local community. According to eyewitnesses:

The girls screamed, threw things about the room, uttered strange sounds, crawled under furniture, and contorted themselves into peculiar positions, ...[10]

Soon after, several other girls suffered similar fits. When the local doctor could find no cause, demonic influence was presumed to be responsible. Several women were arrested on suspicion of witchcraft. These events set off a long period of accusations, trials and executions that spread across the region and lasted until May 1693. By the time the mania was over, more than 200 people had been accused of witchcraft and 20 unfortunates had been executed.

The witch trials of the sixteenth and seventeenth centuries provide an infamous example of people ascribing undesirable events to supernatural causes. The history reveals a tragic confluence off events in which ignorance, prejudice, jealousy and religious fervour all played their part in innocent people (usually older, single women) being condemned to death in reprisal for events in which they played no part. Charles Mackay relates a similar absurd but dreadful incident that occurred in Scotland:

> ... in 1697 a case occurred which equalled in absurdity any of those that signalized the dark reign of King James. A girl named Christiana Shaw, eleven years of age, the daughter of John Shaw of Bargarran, was subject to fits; and being of a spiteful temper, she accused her maid-servant, with whom she had frequent quarrels, of bewitching her. Her story unfortunately was believed. Encouraged to tell all the persecutions of the devil which the maid had sent to torment her, she in the end concocted a romance that involved twenty-one persons. There was no other evidence against them but the fancies of this lying child, and the confessions which pain had extorted from them; but upon this no less than five women were condemned ... they were burned on the Green at Paisley.[11]

Recently, it has been suggested that some cases of so-called "witchcraft" may have resulted purely from ignorance about natural sources of food contamination.[12] Ergot is a mould that contaminates grains of rye. Eating bread made from flour that is contaminated with ergot can lead to a host of nasty results, ranging from vomiting to death. The most striking symptoms are similar to those described in the events at Salem, especially convulsions and incoherent babbling. To fearful witnesses, these bizarre symptoms could easily be interpreted as frenzied dancing and speaking in tongues. Being ignorant of the true cause, they ascribed the symptoms to demonic possession. The unfortunate sufferers thus became victims twice. After suffering the dreadful effect of ergot poisoning, they were put on trial and condemned for witchcraft.

Nevertheless, the tendency to look for simple causes for unwelcome events often leads people to look for the involvement of some human agency. Throughout Europe and North America in the Seventeenth Century, the notion of witches provided a convenient explanation for any misfortune. Mackay described this period of madness thus:

> An epidemic terror seized the nations, no man thought himself secure, either in his person or possessions, from the machinations of the devil and his agents. Every calamity that befell him he attributed to a witch. If a storm arose and blew down his barn, it was witchcraft; if his cattle died of a murrain, if disease fastened upon his limbs, or death entered suddenly and snatched a beloved face from his hearth–they were not visitations of Providence, but the works of some neighbouring hag, whose wretchedness or insanity caused the ignorant to raise their finger and point to her as a witch.[13]

The same phenomenon also occurred in many other settings. A good example is the Biblical story in which a series of events led up to Jonah being swallowed by a whale.[14] When commanded by God to deliver news of his displeasure to Ninevah, Jonah flees from the task and sails off in the opposite direction. When a great storm rises and threatens the ship, the sailors realize that Jonah is to blame. They cast him overboard and the sea immediately becomes calm again. This story underlies an ancient superstition in which sailors will attribute a series of misfortunes to bad luck brought on by the presence of a particular person on board. The name "Jonah" has since entered the language as a term to describe anyone who is seen as a jinx or unwelcome influence.

Although modern research is beginning to make inroads into understanding the links between order and chaos, people still look for simple explanations when things go wrong. One result is the tendency to make a single individual the scape goat for complex events. Accidents, for instance, usually result from processes with many contributing factors, but total blame is often ascribed to a single person who contributed by making a wrong decision at the wrong time. The term "witch hunt" has come to refer to any incident in which people seek out victims to blame for some problem. A notorious modern example was the communist witch hunt during America's McCarthy era. Anyone with a suspect past was cast as a communist, with the implication that they were subversive traitors and spies. The mania extended to guilt by association. Anyone who had contact with a suspect became suspect themselves.

## Here and Now

To sum up the argument in this chapter, people want things to be simple. We like to see simple causes and simple effects. Understanding how things happen, why things happen, is important. If you can understand a thing, then you have the possibility of control. Even if something bad happens, if we can understand it, then it helps us to feel that we have some measure of control.

Unfortunately, the world around us is not simple; it is complex. In the real world, complex interactions and processes are all around us. Long chains of events lead to unexpected events. Often these involve elements that are unobserved unknown to us, beyond our notice, and beyond our control.

Unfortunately people want to ignore complexities. This means that they look for explanations within the confines of what they already know. The problem with this is that our picture of the world and events is biased by the selective nature of perception; our limited, local information; and our individual attitudes and biases.

Harsh realities are difficult to accept. People want "answers," so they seek explanations they can both understand and accept. Unfortunately, complexity means that large-scale patterns of interactions can produce results that locally are unpredictable and capricious. Complex processes are often difficult to identify and difficult to understand, so people tend to look for simple, local explanations. In doing so, they anthropomorphize or assign blame to individuals, rather than systems. Often they blame other people. As we saw above, in the past this tendency has led to witch hunts, bigotry, persecution and untold numbers of wars.

## End Notes

[1]Thomas Schelling's Impossibility Theorem.

[2]There are many accounts of confirmation bias in the psychological literature. For a good review of the many different forms it takes see Nickerson (1998).

[3]The idea first appeared in simulation modelling (Dahl and Nygaard 1966). If you have a model of (say) traffic with lots of drivers, then instead of defining the behaviour of drivers one at a time, an object model allows you to capture features and behaviours that are common to all drivers. The approach was later extended to all kinds of information systems using specialized tools such as the Unified Modeling Language (Larman 2005).

[4]Green (1983).

[5]Estimates of times and population size vary between sources. Most agree that the world's population reached 0.5 billion during the 1500s and 1 billion sometime between 1800 and 1830. The only difference it makes to the model is time at which it projects that population would go off the scale. For an authoritative discussion of the future of world population see United Nations Population Information Network, www.un.org/popin/.

[6]Ulysses, in Shakespeare's *Troilus and Cressida*, Act 1 Scene 3.

[7]Song titled "Fortune, Empress of the World", from *Carmina Burana*, a Thirteenth Century German manuscript, set to music by Carl Orff in 1937.

[8]See Epley et al. (2008) and Waytz et al. (2010). For a short popular account see Fox (2010).

[9]Goodall (1990). The incident related here is described on p. 9.

[10]Wikipedia (2012). Salem witch trials.
   http://en.wikipedia.org/wiki/Salem_witch_trials.

[11]Mackay (1841), p. 542.

[12]Accounts of ergotism can be found in Caporael (1976), Woolf (2000).

[13]Mackay (1841), p. 480.

[14]The full story is related in the Bible's *Book of Jonah*.

# The Animal Within

**5**

*Man is the cruellest animal.*[1]

**Abstract**

Modern humans inherit needs and behaviours from our animal ancestors. These include the desire to belong to a group, the drive for high status within a group and the need for territory and resources. These drives are the seeds of many social problems. Fear of exclusion provides powerful motivation to conform. Leaders exploit this fear, and outside threats, to unite people behind them. The desire for status often leads to destructive competition and self-interested consumption. The inability to cope with traumatic events can lead people into denial. The ancient drive to conquer a hostile and dangerous world still colours people's attitudes towards nature and the environment.

## Basic Instincts

In many respects, we humans are prisoners of our animal past. The primate origins of human nature are well-known.[2] Basic needs and drives often influence people's behaviour and colour their assumptions and decisions. The side effect of being an animal is that our primitive needs and drives sometimes dominate over our long-term best interests. We go on acting on our animal instincts even if they are no longer appropriate.

Instincts, such as an individual's desire for status, can bias people's actions and decisions, over-riding all other considerations. So they can have widespread side effects. Here we will look briefly at the implications of three traits we inherit as social primates: the *need to belong to a group*, the group's *need for territory*, and the *drive for status* within the group.

D. G. Green, *Of Ants and Men*, DOI: 10.1007/978-3-642-55230-4_5,
© Springer-Verlag Berlin Heidelberg 2014

All three of the above traits originated as strategies for survival in the wild. Our ape and hominid ancestors in Africa needed to belong to a group as protection against predators and other dangers. They needed territory to ensure an adequate supply of food. Individuals strove for status to enhance their potential to reproduce.

In civilized society, those primitive traits can be harmful and sometimes dangerous. Unexpected side effects can arise from them. In particular, they motivate the immediacy of self-interest that Merton identified (see Chap. 1).

## The Group

At Kornhausplatz (Granary Place) in the heart of the Swiss city of Bern stands a famous fountain, the Kindlifresserbrunnen, built by Renaissance sculptor Hans Gieng in 1545. What is most remarkable about this fountain is the statue on top, which confronts passers-by with a truly frightful scene. It portrays a huge ogre in the act of eating a child. At his side is a sack full of captured children yet to be devoured. The purpose of this gruesome monument, locals informed me, was to scare disobedient children. It showed them, in the most graphic terms, the fate that awaited them if they strayed away from the safety of their home and family.

The drive to be part of a group stems from the need to survive in the wild. For herd animals it is an effective defence against predators. Hunters who operate as a team are far more effective than one hunter working alone. Driven by such advantages, humans have inherited a basic need to be part of a social group.

Pointing out the danger of straying away from the safety of the group  is one of the oldest ways to maintain group discipline. Given that safety lies within the social group, fear of exclusion is one of the greatest motivations for conforming. In the medieval Church, excommunication carried the implication of exclusion from all civilized society. One of the harshest punishments in ancient Rome was proscription. Not only did victims lose all their property and possessions, but also any of the usual protections that society provided. Anyone could rob or kill a proscribed person without fear of punishment.

One trend throughout history has been the increasing scale of group identity. Family groups gave way to tribes, tribes to nations, nations to empires and lastly, one would expect, empires would give way to world government. Unfortunately this has not happened; at least not yet. The United Nations ought to fill that role, but its members are too caught up by their own, local self-interests to give control to what they still regard as outsiders.

The anthropologist Robin Dunbar argued that primate societies have a natural group size.[3] Groups that grow too large spontaneously break apart into smaller groups. For baboons and other apes the natural group size is 30–60 animals. Grooming is an important element in keeping the social group together. Dunbar argued that in humans, gossip (social talk) replaces grooming and is more efficient: one person can talk to several others all at once. He cites a variety of evidence to support his claim that the natural group size in humans is about 100 to 150 individuals. In the Roman legions, for instance, the basic unit was the Century. He also cites the leaders of some religious sects in the United States who have found that when the size of a community grows to more than 150 people, it becomes impossible to maintain order by peer group pressure alone. When communities grow larger than this limit, some people must break away and form a new colony.

Large social groups maintain order by combining peer influence, which provides pressure from the bottom up, and prescription (e.g. laws and rules), which provides pressure from the top down.[4] They are usually formed around central symbols or causes that people share in common, such as a place, a flag or a belief that gives the group an identity and purpose. The weaker the symbol, the more difficult it is to maintain social consensus and cohesion. Conversely, the stronger the central cause, the less top-down pressure leaders need apply.

In the early 1980s, Margaret Thatcher, Britain's first female prime minister, was trailing behind the opposition in opinion polls. It seemed almost certain that she would be voted out of office at the next general election. Then, on Friday 2 April 1982, Argentina invaded the Falkland Islands, a remote island territory in the south Atlantic. Thatcher promptly sent a naval task force, which recaptured the islands after a war lasting just two months.[5] Riding a wave of national pride, Thatcher leapt ahead in the opinion polls. She won the 1983 general election in a landslide and became one of Britain's longest serving prime ministers.

From the beginning of recorded history, and no doubt long before, leaders have known the value of having an outside enemy to unite people behind them. Perhaps one reason why a global group identity has not taken hold is that there are no Martians waiting out there for humanity to cast in the role of the demon enemy.

More generally, there is nothing like the fear of losing something to make people cling to what they have. This fear tends to make people more conservative as they grow older and as they acquire wealth and property. The greater the threat, the more fiercely people rally around their central interests. And they rally to groups that will best support those interests. In Japan, for instance, loyalty to the clan in historic times gave way to loyalty to the company.

Leaders often exploit fear of loss to persuade their followers. When there is no obvious outside threat, leaders often try to create one. The threat may be external, or internal. This often gets combined with a tendency to demonize a different group to enhance the identity and solidarity of some outside group. Hitler infamously tried to blame his country's ills on the Jewish community.

## No Place Like Home

In December 1968 the Apollo 8 spacecraft became the first manned flight to leave Earth orbit. As it passed around the Moon the crew witnessed something no one had ever seen before: the sight of the Earth rising in the sky before them. For the first time people were able to see our entire planet as it appeared from another world.

Historically, group identity is often coupled with a location. In many species, animal groups have home territories, which they defend against encroachment by other groups. Humans do the same, at all levels, from neighbours arguing over the boundary lines, to nations battling over disputed islands. Beginning with Robert Ardrey[2] in the 1960s, many authors have examined both the origins of territoriality in nature and its effects on human behaviour.

Historically, of course, most human wars have been fought over territory, and usually over the resources that the land had to offer. This continues to be the case even today. During the 2003 invasion of Iraq, much argued claims about weapons of mass destruction overshadowed the more basic motivation of ensuring continuing oil supplies.

The identification of people with the land they occupy can be very strong. It seems ironic that while globalization sees the worldwide spread of organizations and increasing mutual interdependence between peoples in far distant parts of the world, the attachment of many societies to their land has grown stronger.[6]

In the wild, territory is associated with an animal's group, with its food supply, and with safety. For modern humans, territory takes on new forms, including identity, income and security. In the professional world, territorial issues sometimes arise around areas of responsibility and the demarcation of duties. In extremes, this can lead to the formation of fiefdoms within an organization, each jealously guarding its patch.

People will guard their patch even when it is counterproductive. In the early 1990s, I was involved in several projects that aimed to bring together data from many sources so the authorities could better manage environmental conservation. The technical challenges of these projects were huge, but they paled in comparison to the human challenge of convincing people to cooperate. Instead of seeing the good of the country, all they saw was an invasion of their professional space. No one was willing to share their hard won data. To them data not only provided a source of income, but also embodied their expertise and authority. As one botanist complained: "If they want to know where to find a plant, they should come and talk to me." In another case a government agency wanted to charge such a high price on their data, the cost was more than our entire budget. Fortunately higher authorities stepped in when I pointed out that this meant the government could not afford to use its own data.

## Looking Good

The desire for status is a powerful social drive. It influences many decisions and actions. To maintain status, people feel they need to conform to social trends. The entire fashion industry is based on the desire for status. Encouraged by commercial image-mongers, the desire for status also leads to conspicuous consumption. People crave a big house in an exclusive neighbourhood, an expensive car, designer clothes and all the latest gadgets and accessories. By all these means and many more people hope to prove that they have high status, even if they go deeply into debt to do it.

Like it or not, we humans inherit a pack mentality from our distant ancestors. Along with this mentality, we have inherited dominance hierarchies and struggles for status. Animals assert their dominance over pack members by aggressively putting down any behaviour that draws their status into question. Unfortunately, humans have a tendency to do the same.

It is a sad fact of life that one of the easiest ways to look good is to make someone else look bad. One of the worst examples of this I ever saw was at a workshop where a senior colleague systematically bullied and humiliated his juniors, most of whom were making their first formal presentations about their work. His technique was simple: ignore anything positive in the presentation and latch onto some issue the speakers had not dealt with (no matter how trivial) and treat it as a gaping flaw that invalidated everything they had done.

Many of our institutions and practices have evolved to provide platforms for people or groups to make themselves look good by making someone else look bad. For example, our legal system is structured around "legal battles" between opponents. Lawyers seek to make their case by picking legal holes in their opponents' arguments.

The Westminster style of parliament has a government and an opposition. Typically the opposition tries to make itself look good by attacking and criticising every motion put forward by the government. This means that parliamentary debate usually reduces to a test of strength. Delegates on the opposition benches automatically attack any proposal put forward by the government. The government, in response, automatically defends its position on every point. Having a greater number of delegates, the government normally wins and their motions are passed. The opposition often raises valid issue that should be addressed, but there is no room for compromise. The government feels compelled to defend itself on every point, so flawed proposals are often passed without correction.

The same thing happens in boardrooms. Arguments often fail to reach the best conclusion because they become contests for status. To the individuals involved it may seem more important to win the argument, rather than find the best answer. When someone proposes a new idea, ambitious members of the board may treat it as an opportunity to promote themselves. Instead of looking for ways to develop and apply an idea proposed by a rival, they are more likely to look for flaws they can use to criticize and demolish.

In the workplace, this phenomenon manifests itself in many other ways. Finding fault with colleagues is common. A boss may habitually criticise his/her subordinates as a means of reinforcing superiority. Some workers become sticklers for rules, having found that it provides them with endless opportunities to score petty victories by pointing out errors in procedure by their colleagues.

Another variation on the above theme is to blame others to avoid looking bad yourself.

Boss:       *"Alright, we've just lost the Franklin account. Someone failed to deliver the goods to them on time."*
Fred:       *"It's George's fault. He's in charge of customer deliveries."*
George:     *"Rubbish. Fred didn't tell me they needed it today."*

Who is really at fault? It could be Fred or it could be George. Maybe both are partly to blame. Maybe it is neither of them. Whatever the truth, they each try to avoid looking bad by blaming the other.

The same thing happens at a group level. An entire group of people seeks to deny fault for their mutual failings by finding scapegoats. As we saw earlier, Hitler made the Jews scapegoats for Germany's economic problems during the late 1920s and early 1930s. In the 1700s, French revolutionaries made Marie Antoinette (a foreigner) the scapegoat for economic woes that really arose from a long series of wars, bad economic management, and the greed of the entire nobility.

If making others look bad can make you look good, then the reverse occurs too. By making themselves look good, people sometimes (whether by accident or design) make other people look bad. A new employee joins a company and being young, eager and hard-working, she achieves much greater productivity than her colleagues. Far from being pleased for her contribution, the older workers resent her for making them look bad.

## Elvis Lives

The death of Elvis Presley, the "King of Rock", in 1977 was a shock to his many fans. For some of them, the shock was so great they simply refused to believe that he was really dead. To account for his disappearance, a story spread that Elvis had simply gone into hiding. Fans rationalized this idea by suggesting that he retired from public view to escape constant public attention.

Harsh facts are difficult to accept. Sometimes they are so hard to accept that people need a long time to come to terms with them. Through her observations of cancer sufferers, Elisabeth Kubler–Ross was able to identify five stages of grief that people go through when reacting to devastating news[7]:

Stage 1    *Denial* (This cannot happen to me!);
Stage 2    *Anger* (Why is this happening to me?);
Stage 3    *Bargaining* (I'll do better if only …);
Stage 4    *Depression* (I don't care anymore);
Stage 5    *Acceptance* (I'll take whatever comes).

The above stages are not confined to people who learn they have cancer. Devastating news takes many forms, such as hearing about the death of a loved one, the loss of a long-held job, or financial ruin. Really the above stages of grieving could apply to anyone who experiences anything that shocks them to the core. Indeed, given this problem of coping with bad news, it seems possible that people may respond in a similar way to any unpalatable change in their lives.

The Holmes and Rahe Stress Scale measures the degree of stress associated with traumatic life events.[8] In a study of 5,000 medical records, the authors found how closely related different events are with illness. Their scale rates common events on a scale from 0 to 100. Among the most serious are events such as death of a spouse (100), divorce (73), imprisonment (63), marriage (50) and dismissal from work (47). Taken over a year, a total score greater than 300 identifies people at risk of illness, while anyone with a score less than 150 has low risk.

Alvin Toffler[9] described a phenomenon he called "Future Shock," which occurs when people experience too much change happening too quickly. They are simply unable to cope and adapt to massive disruption in their lives. He argues that this future shock is a widespread phenomenon in western society today. There is a lot of evidence to support his claim. Many authors have pointed to the way the world seems to be speeding up, moving faster and faster.[10] Also, the rate at which new technologies are being introduced has become a virtual flood (see Chap. 12), so technological change is producing more widespread social side effects than ever before.

Finding it difficult to cope with huge, unpalatable changes, some people respond in a way that is similar to coping with grief. Sometimes people never get past the first stage above. They do not get past the denial stage.

Persistent denial of harsh realities can lead to strange results. In the case of Elvis, rumours of his continued existence provide an amusing but harmless aberration amongst some of his fans. But when the harsh facts concern issues of public importance the results can have serious consequences.

The difference between a skeptic and a denialist lies in how they deal with evidence.[11] A skeptic asks questions and follows wherever the evidence leads. In contrast, a denialist filters the evidence, picking out evidence that supports his/her prior belief. At the same time he or she ignores (or if need be casts aspersions on) evidence that refutes that belief. This is a case of confirmation bias,[12] which we met in the previous chapter.

Diethelm and McKee documented cases where reaction to unacceptable truths led to public movements that thwarted efforts to deal with major issues, such as global climate change.[13] They found that to achieve their aims, "denial movements" (to given them a name) all employ the following six steps proposed by Hoofnagle[14]:

Step 1 *Allege a conspiracy.*
Claim that scientific evidence is collusion not facts.
Step 2 *Use false evidence.*
One approach is to provide pseudo experts. These are people who claim to be experts in the particular field, but espouse views that are contrary to well established knowledge.
Step 3 *Cherry pick the evidence.*
Pick out the bits that agree, ignore contrary evidence. Cherry picking evidence leads to confirmation bias. People seize on every piece of evidence that reinforces or confirms their opinion and ignore evidence that disagrees with them. The more firmly they hold an opinion, the more they selectively filter evidence. For instance, highlight any flaws in studies and use these to question the entire field. Creationists, for instance, repeatedly point to gaps in the fossil record, claiming they are fatal flaws in evolutionary theory.

Step 4 *Create impossible standards for opponents.*

In scientific research, especially in biology and medicine, random factors often make 100 % results impossible. A new vaccine may cure most peoples, but not necessarily all. To cope with this problem, science uses statistics as a guide to truth. So a statistical test might support a hypothesis with, say 99 % confidence. But this means there is a 1 % chance that the hypothesis is incorrect. This tiny lingering doubt provides a crack for denialists to argue that the case was not proved.

Step 5 *Use logical fallacies.*

This includes the use of *red herrings, straw men, false analogies,* and the *excluded middle* fallacy. For instance, an example of the excluded middle might run like this: "Hitler opposed smoking, so anyone who opposes smoking is a Nazi." This false reasoning excludes the rather large middle ground: people who oppose smoking but are not Nazis.

Step 6 *Manufacture doubt.*

A little misinformation goes a long way. If there are two opposing views, then an uninformed public will not be able to tell which side is correct. The cry "both sides must be heard" can sometimes lend undue credibility to malicious lies. Thus the media aim of giving a "balanced account" sometimes leads to a totally imbalanced account.

## Conquering Nature?

The history of science over the past 500 years is a story of slowly increasing humility in the face of nature. Prior to Copernicus anyone would have said that the sun, the stars and all the planets revolved about the Earth. Humankind was the peak of creation, created in God's own image.

Like young children, the human race, and especially those of European extraction, considered that they were the centre of the universe, made in the image of God. This belief underlies an assumption that everything on Earth was provided for humans to use as they please. The natural world was seen as vast, mysterious and full of danger. Even in the late Twentieth Century, geographic exploration and settlement were still often portrayed as "conquering nature."

One after another, scientific discoveries have gradually peeled away humankind's hubris. It began with Copernicus showing that the Earth was not the centre of the universe. The sun does not revolve around the earth; we live on just one of many planets revolving around the sun.

Later astronomical discoveries downgraded our place in the universe even further. Not only were we not at the centre, our planet circles a medium-sized star at the very fringes of a galaxy containing billions of stars. And that medium-sized galaxy is likewise just one of billions visible within the known universe.

It has taken humankind centuries to recognise that we are not the ultimate end of creation, made in God's image at the centre of the universe and free to loot and pillage the world's resources to satisfy our every whim. However, there is still great reluctance to admit that we are subject to the same rules that govern the rest of nature.

During the Nineteenth Century, the theory of natural selection provided another shock by showing that, far from being above and apart from nature, humans, chimpanzees and other apes all evolved from a common ape ancestor that lived some 13 million years ago.[15]

Despite all that has been discovered, many scientists still find it hard to come to terms with humankind's relationship to the rest of the animal kingdom. This is evident in dogged rearguard attempts to find scientific proof of humanity's uniqueness. The implicit argument would run something like this: "We may have evolved from apes, but we are still the pinnacle of evolution. Lots of things separate us from the rest of the animals, making us both superior and unique."

However, during the Twentieth Century, a series of discoveries in biology steadily chipped away at this assumption that humans hold a unique place in nature. One assumption was that everything about humans—our genes, our brains, our culture and our society—was either unique or more complex than in other life forms. For instance, when scientists began to sequence DNA, it was widely assumed that the human genome would turn out to be one of the most complex of all. With a total length of about 3.08 billion base pairs, human DNA certainly far exceeds the size of DNA for common viruses (at most a few thousand base pairs). But it was something of a surprise to find that humans have a much smaller genome than many "lower" organisms, such as grasshoppers, peas and even lilies.[16] Also, it turns out that the genomes of humans and chimps differ by only 3 % of their genes and only 1.1 % of their DNA sequences.

People have always assumed that humans are more intelligent than any other animal. But what exactly does this mean? How do you measure intelligence in a way that allows you to compare humans and animals?

The problem is compounded because it is hard to pin down exactly what defines intelligence.[17] Alan Turing, a computing pioneer, offered the following heuristic test for intelligence in computers: if a computer can carry on a conversation with a human and the human cannot detect that it is a computer, then that computer is intelligent.[18] In the late 1960s, a program called ELIZA revealed the flaw in Turing's test by showing that much of human conversation involved almost no intelligence at all.[19] What ELIZA did was to use a script of stock phrases (e.g. "Why do you say that?") and simple rules that threw bits of a user's inputs back at them. Designed to mimic aspects of psychiatric dialogue, ELIZA was so captivating that some people became hooked on it.

One simple measure of animal intelligence is brain size. A human brain is bigger than a dog's, so humans are more intelligent, right?[20] The trouble is that some animals (such as elephants and whales) have brains that are bigger than humans. It was argued that these animals have large brains only because they need to process sensory data from much larger bodies. Using the ratio of brain to body weight took care of the elephants, but humans still did not come out on top; even the mouse scored better. To deal with this new problem, they played with statistics and came up with a new measure, called an *encephalization quotient* (EQ) that did put humans on top.[21] But other animals, especially dolphins, have high values too. The matter continues to be debated.

The above problem raised the question of what dolphins are actually doing with such large brains. The answer (at least in part) it seems is that dolphins need large brains to process the 3D images produced by their sonar clicks. But are they intelligent?

Given the problems with measuring intelligence, a safer line of argument seemed to be that human intelligence differs not just in quantity, but also in quality. What differentiates humans from animals? One claim was that only humans use tools. However, it was soon shown that many animals not only use tools, but in some cases they even make them.[22]

Well what about culture? Only humans pass on new cultural discoveries, surely? This too failed. Recent studies of macaques, for example, have confirmed that monkeys not only pass on tool-using skills from one to another, but also adapt those tools to other uses.[23] Also the infant macaques are most likely acquiring tool using ability from their mothers.[24]

A lot of attention has centred on language. Although animals communicate with one another, they do not make up sentences or convey abstract ideas. Much research focussed on trying to detect evidence of language in dolphins and whales, but without success. More successful have been experiments that try to teach dolphins, chimps and gorillas some version of simple human language. Apes do not have the vocal apparatus for speech, but are able to learn sign language and use it to communicate with their trainers and other chimps. Critics argue that these chimps have not really mastered grammar but merely learned the meaning of particular combinations of signs.

The problem in these language studies was that critics kept moving the goalposts. Each time research demonstrated that animals are capable of some feat ascribed to humans alone, the critics would raise a more subtle issue as the defining character of language. In essence these objections boiled down to defining language as the way we humans communicate. This raises again the question of how dolphins communicate. They seem to be able to pass on complex information from one to another. Is it possible that their way of communicating is completely different, but no less complex and subtle than human language? Using echolocation, for instance they are capable of perceiving 3D features, and use sonic whistles for communication.[25]

A perversely negative approach to proving our uniqueness has been to point to the nastier deeds that people do. Only humans, it was argued, carry competition to extremes such as murder. However, research by Jane Goodall showed that even our darker side is not unique. Chimpanzees for instance, sometimes indulge not only in murder, but also cannibalism and warfare.[26]

In short it seems there are no real qualitative differences between humans and other animals. Admitting our animal nature is important. For hundreds of years, especially since the industrial revolution, humans have seen themselves as apart from nature. On the one hand, there were plants and animals and on the other hand there were humans. Humans observed the natural world but did not see themselves as part of it.

Assumptions about humanity's place in nature have had immediate effects on the ways in which people treat animals, plants, and the environment in general. If we are created in God's image and living in a world created specially for us, then we are justified in exploiting the world's resources to the full. On the other hand, if we are part of nature, and mutually dependent with it, then we need to treat it with greater care and respect. We should not try to conquer nature, but to conserve it.

Historically, people have regarded nature as mysterious and dangerous. Primitive tribes used to send young men out alone into the wild as a test of worthiness. So while there remained new lands to discover, mountains yet to climb, the poles to reach and ocean depths to explore, people remained obsessed with conquering nature. Challenging nature in any way is still seen as both a test of bravery and a test of human achievement.

For over 500 years, western civilization has been conquering nature, gradually pushing back the frontiers of the unknown. First there was the great age of exploration. Navigators such as Columbus and Cook gradually filled in blank spaces on the world map. By the Twentieth Century, exploration had become a matter of reaching the extremes: Admiral Peary reached the North Pole in 1909; Amundsen reached the South Pole in 1911. Tenzing and Hillary scaled Mount Everest, the highest point on Earth in 1953 and Walsh and Pickard descended to the depths of the Mariana Trench in 1960. But the conquest of nature goes on. Having "conquered" the Earth, humanity turned its attention to the stars, landing on the Moon in 1969. By the end of the Twentieth Century, humans had sent robot probes to visit every planet in the Solar System and even beyond.

When you have exhausted all the challenges of the unknown, when you control every part of it, your perception of a thing changes. No longer is there anything mysterious and threatening "out there". The challenge moves from conquering to conserving. How should we best use the world we have conquered?

Denying our animal nature is to overlook issues that often motivate people's opinions, decisions and actions. As this chapter has shown, we need to understand that we inherit certain traits from our animal nature and, even more importantly, we need to be aware of their side effects. Are we the cruellest animal?

---

## End Notes

[1] Ascribed to Friedrich Nietzsche.
[2] For example Ardrey (1966), Desmond Morris (1967).
[3] Dunbar (1996).
[4] See also the section *Law and Disorder* in Chap. 15.
[5] Freedman (2005).
[6] Freidman (1999).
[7] Kübler-Ross (1973).
[8] Holmes and Rahe (1967).
[9] Toffler (1970).
[10] Gleick (1999).

[11]Shermer (2010).

[12]There are many research publication about confirmation bias. Widely cited reviews include Klayman and Young-won (1987); Nickerson (1998).

[13]Diethelm and McKee (2009).

[14]Hoofnagle (2007).

[15]Different lines of evidence have suggested dates for human-chimpanzee divergence that range from 6 million years to 25 million years. The date of 13 million years is the current best guess. For more details see White et al. (2009).

[16]The human genome has about 3.08 billion base pairs (bbp), which includes about 26,000 functional genes. Many species have larger genomes, for example the pea has $\sim 4$ bbp, the grasshopper 13.4 bbp and the lily 124.8 bbp. See National Center for Biological Information: www.ncbi.nlm.nih.gov.

[17]See Chap. 9 for a discussion of the notoriously difficult problems raised by attempts to measure human intelligence.

[18]This idea provided the basis for Philip K. Dick's famous sci-fi novel *Blade Runner*.

[19]Weizenbaum (1966).

[20]Note that in humans, brain size has very little to do with intelligence. According to the Guinness Book of Records, the largest brain ever recorded belonged to Ivan Turgenev, a famous Russian author, and the smallest brain ever recorded belonged to Anatol France, a famous French author.

[21]The EQ is obtained by first calculating brain/body weight ratios for a group of related species (notably the mammals). Using this distribution, the EQ for each species was defined by the "observed" and "expected" values. One use has been to study brain the capacity of extinct species from fossils.

[22]For example Blue Jays have been observed to make simple tools (Jones and Kamil 1973).

[23]Macellini et al. (2012).

[24]Huffman et al. (2008).

[25]The precise nature of dolphin communication is the subject of many studies. See for instance, Herman and Pack (1993), Janik (2000), Harley et al. (2003), Gregg (2013).

[26]Goodall (1999).

## 6

# More Things in Heaven and Earth

*There are more things in heaven and earth, Horatio,*
*Than are dreamt of in your philosophy.*[1]

**Abstract**
Many problems arise from limited thinking. Our mental models—the ways we think about the world around us—are limited by our experience. Gaps in these models mean that unexpected conditions often arise. When confronted with conditions we have never encountered before, our models fail. Accidents and other disasters may follow. A contributing factor is that we are trained to limit our thinking in ways that society expects; another is the tendency to think of situations as a closed box, which rests on the assumption, often false, that the outside has no effect at all. Examples of the resulting problems include the tactics that led to slaughter on the battlefields of World War I, problems with America's first space flight, and the fatal assumption of the unsinkable Titanic.

## Over the Top

At 7:30 a.m. on the morning of July 1st, 1916 thousands of British and French troops swarmed out of their trenches along a 40 km front and began an assault on the opposing German lines.[2] Their advance marked the start of the Battle of the Somme, which was to prove one of World War I's deadliest battles.

Within seconds the bodies began piling up. Many soldiers never made it out of their trenches: as they rose to go over the top, they were mown down by machine gun fire. By the end of that first day, over 19,000 British soldiers lay dead and another 38,000 more wounded or missing. It was slaughter, the greatest slaughter in the history of the British army. All the other countries involved suffered too.

D. G. Green, *Of Ants and Men*, DOI: 10.1007/978-3-642-55230-4_6,
© Springer-Verlag Berlin Heidelberg 2014

By the end of the battle, the casualties from both sides totalled over a million, including more than 300,000 dead.

This incredible slaughter took place because the nature of war had changed. Advances in military technology had outpaced military thinking. The generals based their tactics largely on traditional ideas, on military principles that dated back to Napoleonic times. But the assumptions underlying those tactics were no longer valid. The central idea was that you overwhelmed an enemy's defences. You lined up your men and marched them at the enemy, counting on sheer numbers to break through. Even against muskets this tactic had been wasteful of lives. Against machine guns and modern artillery it was sheer murder. Wave after wave of men were lined up, sometimes carrying heavy loads of equipment, and ordered to walk towards the enemy lines. Almost to a man they were gunned down within a few yards.

The Battle of the Somme was not the first episode of mass slaughter during the war, and it was far from being the last. It took several years, and the loss of millions of lives, before military leaders finally learned that they had to abandon traditional thinking and discovered how to mount effective attacks. The Great War, as it was known at the time, became a war of attrition. Technological innovations, such as poisonous gas, armoured tanks and aeroplanes, eventually played a part. In the end, however, an armistice was called because Germany was simply too weakened to continue.

## Reality Unchecked

Our lives are built on assumptions. We assume that a chair is solid and sit on it without thinking. We assume that the food we order at a restaurant is clean and eat it without a second thought. Assumptions like these are essential. If we tried to question everything, everywhere, all the time, then life would be impossible. Try to guard against every possible source of infection, for example, and you end up like the paranoid billionaire Howard Hughes, a prisoner in his hotel penthouse.[3]

On the other hand, we become so used to making convenient assumptions that we often fail to ask questions when we should. If the chair is not solid, then we sit on it and crash to the floor. If the restaurant has not reached health department standards, we eat and suffer the painful consequences.

In general, problems arise when our assumptions fail to match reality. Some people hang on to assumptions rather than face unpalatable alternatives. Others hang on to old assumptions because they are overwhelmed by change and simply do not know how to cope.

The generals in World War I had never learned how to deal with the greatly increased firepower on the Western Front. In the absence of realistic solutions, they fell back on old ideas and assumptions. The French, for instance, decided that all they had to do was to update their old tactics. The machine gun fired faster than a rifle, so what was needed, they decided, was to be more aggressive. At the start of the war, they drilled into their officers the idea of élan ("dash"). As a result, the French army suffered over 600,000 casualties in the opening three weeks of the war.[4, 5]

As we saw in earlier chapters, underlying our assumptions and actions are schemas, mental models of the world around us. From model ships to fashion parades, from portraits to road maps, models of one kind or another are every-where in the modern world. Most importantly, we create models of the world around us inside our minds.

Regardless of what form they take, the important thing to know about models is that they are never complete. As we saw in Chap. 4, they leave things out. And it is the things they leave out that jump up unexpectedly and give us grief. That models have limitations is well known to scientists. But it is important to realize that our mental models are limited too. In other words, they often fail. Just such a problem arose when America was preparing for its first manned space flight.

## The Space Race Springs a Leak

In 1961, the United States of America was engaged in a space race with the Soviet Union. A lot depended on the outcome: commercial success, international prestige and military advantage all hinged on dominating the "high ground" of space. But America was losing. In April 1961, Russia launched Yuri Gagarin into orbit, winning the race to put the first man into space. America, on the other hand, had rockets that had a disturbing history of blowing up during launch.

All of this made it supremely important that America's first attempt to launch a man into space should go off without a hitch. So it was that on May 5th 1961, astronaut Alan B. Shepard found himself strapped inside a Mercury capsule on top of a Redstone rocket, waiting to become America's first man in space.[6] Given the importance of the flight, and their history of recent failures, it was no surprise that the launch control team were carefully checking and rechecking every aspect of the rocket and its control systems.

A manned rocket is an extremely complex object. The Redstone missile, which NASA had adapted for Shepard's flight, contained thousands of parts. There were hundreds of systems and monitors that needed checking, ranging from fuel and guidance control to power levels inside the manned Mercury capsule. Shepard, a former naval aviator, had dozens of sensors taped to his body to monitor his condition and to download records for research into space medicine.

The Redstone rocket did not have enough power, or fuel to launch the capsule into Earth orbit. Instead, this first flight was to follow a simple ballistic trajectory up into space and straight down again. The whole flight would last only 16 min. In truth it was largely a public relations affair to retrieve some credibility for the American space programme, which up until that point had enjoyed few successes.

The technicians were under enormous pressure to get it right. In the previous three weeks, two rockets went off course during tests of the Mercury system and had to be blown up. For this to happen with a man on board would mean international humiliation. Everyone in launch control knew that their reputations, not to say their jobs, were on the line. They had to get it right. No one wanted to be the one who fouled up.

Technical problems, especially an overheated inverter, had led to a 4 h hold in the countdown. Worried about other potential problems, the launch controllers were checking and rechecking every system just to be absolutely sure they had not missed something.

As it turned out they *had* missed something. All these checks and rechecks were taking time, lots of time. Although the flight was due to last only 16 min, the pre launch procedures had taken 4 h. By this time the coffee that Shepard had drunk at breakfast that morning had passed through his system. He was bursting. He needed to pee.

When Shepard reported his problem, it created an unexpected dilemma for the launch controllers. Either they remove the astronaut and postpone the launch, or let him pee in his suit. Shepard asked permission to relieve himself inside his space suit. However, this would have unpredictable consequences. It could lead to a short circuit in the sensors strapped to the astronaut's body. In the pure oxygen environment of the Mercury capsule, the risk of fire from a short circuit was extreme. Six years later, three astronauts were killed on the launch pad when fire raged through their capsule during a test.

The alternative possibility was to reschedule the launch for another day. With millions watching, such a delay would be highly embarrassing. Reluctantly, mission controllers opted to let Shepard pee in his suit. Immediately, his bio-medical monitors went crazy and alarm bells started ringing in the control room.

When controllers still dithered with the countdown on hold at T minus 2 min and 40 s, the astronaut himself forced their hand.

"All right, I'm cooler than you are." Shepard challenged them. "Why don't you fix your little problem and light this candle."

Worried that something else might go wrong, the flight controller finally gave the go ahead. To everyone's relief, the launch went off without a hitch.

You think you have things under control. You have planned and checked and taken into account every conceivable possibility. Then something unexpected, something inconceivable, crops up and wreaks havoc. The engineers who designed and planned the Mercury space mission thought they had a complete model of the rocket and its systems. They thought they had taken everything into account. They had not. As we saw above, their model overlooked a crucial detail—the astronaut. Fortunately for them, the oversight was not catastrophic.

## Closing the Box

One of the most common, and most damaging, ways in which people's models fail is by treating a problem as a closed box. This is what led to the astronaut's embarrassing problem in the above story.

When engineers build a bridge, they consider all the relevant technical issues that fall within their expertise. They deal only with the immediate problem of building the bridge. They are not usually asked to consider all the other issues involved, such as the impact the bridge might have on people living in the neighbourhood. Such side effects are the responsibility of the planners who decide that a bridge is needed.

The problem is that all too often, the wider implications of events are overlooked. In many cases, they are virtually unpredictable. In his 1996 book *Why Things Bite Back*, Edward Tenner made the point that every technological innovation has side effects.[7] Sometimes these side effects create problems that are more severe than the problem they solve. We will look at some of these problems in later chapters.

The closed box view of the world lends itself to problems that arise, seemingly out of the blue. But in reality they occur because people have oversimplified the problem by ignoring the wide context in which it occurs.

One of the underlying problems is that society conditions us to limit our thinking. Children are trained to think in the ways society expects. The intent is to make them grow into good citizens. So patterns of thought that stray outside the norm are discouraged. Unfortunately, this includes creativity and seeing complex patterns. They are taught to think along simple lines. This unconscious agenda is evident in intelligence testing: certain types of questions rely on identifying the "correct" pattern or classification. As an example, look closely at the following question, which is one kind of question found in tests to assess intelligence.

QUESTION Carrot is to banana as potato is to …
  (a) basketball (b) pinecone (c) celery

Presumably the testers expected children to realize that carrots and bananas are both types of food, so the "correct" answer is celery. But a child could equally well notice that carrots and bananas have similar shape and therefore select basketball because it has a similar shape to a potato. Being even smarter, a really observant child might notice that a carrot and potato are both roots of plants, but a banana is a seed-bearing fruit. So they would select pinecone as the correct answer. It is easy to dismiss this particular example as being just a bad question, but problems of this kind lurk in many questions I have seen in IQ tests.

Test questions such as the above reward thinking within the box and punish the creative thinker who looks for fresh associations between things. The problem is that any of the above answers is "correct"; it just depends on what model you choose to use when looking at the world. In effect, psychologists who design IQ tests are saying that only one model is correct; only one worldview is acceptable.

The problem is that thinking inside the box can lead to unexpected consequences in social interactions between people. It is said that communication breakdowns account for ninety percent of interpersonal problems. One reason for this is that each person has their own mental models of situations and relationships. This means that when any two people interact with one another, they will interpret the experience in different ways. And these differences can lead to problems. Consider the following examples.

*Case 1.* A child is playing while his mother is cooking in the kitchen. Curious about what his mother is doing, he wanders into the kitchen and climbs up on a stool to look into a pot of water boiling on the stove. His mother pulls him out of the kitchen and gives him a smack. The way she sees it, she has just taught him to stay away from dangerous things. However, the way the child sees it, climbing up to the stove is a sure way of getting attention from his mother.

*Case 2.* Sally is a medical student preparing for her exams. Knowing how busy she is, her boyfriend George thinks he is doing her a favour by keeping out of her way so she can focus on her all-important studies. Sally, on the other hand, is stressed about the exams and desperately wants to see George's friendly face. When he doesn't ring, she begins to think he is seeing someone else. When the exams are over and they finally do meet again, their frustration boils over into a blazing row that threatens their relationship.

## More Things in Heaven and Earth

In one way or another, our lives are surrounded by models. How often, for instance, have you seen charts showing how a large company is organized? Typically it will be a pyramid of some kind, with the CEO at the top and divisions dealing with finance, sales and so on. This is a model, a model of how the company works.

Likewise, the road rules are built on a model of how traffic should flow. They assume that drivers should stick to one side of the road. The rules are there to constrain drivers so that they do stick to one side of the road, so that they do behave as the model says.

So, whether it is a car we drive, a company we manage, or a space flight we plan, we base our actions on models. And because the real systems are more complex than our models, there are always missing details and unexpected situations.

System designers know about missing details. It is sometimes called the *platypus effect.* An unexpected feature or situation crops up and forces you to go back to the drawing board (sometimes literally) and redesign your model. The term derives from the predicament faced by science when they first encountered the platypus.

In 1798, scientists at the British Museum in London received a specimen of a weird animal from the remote colony of New South Wales. The skin appeared to represent an animal that defied all their preconceptions about the animal kingdom. It had fur like a mammal, and suckled its young. And yet it had a bill like a duck and laid eggs like a reptile. Not surprisingly, the scientists were deeply suspicious. The creature defied every rule of taxonomy. Either the whole thing was a hoax or they had to revise all their theories about animal taxonomy. Convinced that the colonists were playing a joke, they reasoned that the specimen had been sewn together using parts of different animals. To this day, the specimen still retains marks left by a scalpel used to probe its skin, looking for stitches. It was only when they saw a live specimen of the animal, which we now know as the platypus, that they reluctantly admitted it was real. Their reluctance stemmed from the unpalatable fact that it forced them to revise all their assumptions about distinctions between groups in the animal kingdom. Today we know that the platypus is a member of the monotremes, a group of primitive mammals that retains some features (such as egg-laying) of its reptilian ancestors.

The same kind of mistake occurs in everyday life. We assume that our local knowledge of events around us represents all there is to know. This assumption leads to surprises when situations arise that you have never met before. It is particularly common when people think in terms of simple cause and effect.

One area that is notorious for having to cope with inexact models is the law. The legal system is forever patching up laws by adding new rules and creating precedents. This effort is necessary because the underlying ethical models are imperfect, and because language is never able to capture everything.[8]

In effect ethical values form a model of how people should behave and how they should treat each other. Just like the road rules, laws serve as constraints to force people to behave in an ethical manner. Laws are made to confine people's behaviour to what is acceptable within that model.

However, there is a problem. As everyone knows, the law is far from perfect. Like all models, it leaves out many things. So there will always be new circumstances and odd situations that never occurred to the lawmakers. These omissions make for exceptions, contradictions and loopholes that the unscrupulous can exploit. Also there can be instances where ethical values conflict. Which is worse: to let a child starve or steal a loaf of bread?

The result is that the law is continually being patched up. Judges make decisions about grey areas and these decisions create precedents for future cases. Over time the precedents pile up, forming a mountain of detail far beyond the ken of ordinary citizens. This accumulation of legal detail leads to a cascade of consequences, such as expanding bureaucracy, specialized legal departments, crippling legal costs, and abuse of legal technicalities to bypass restrictions or crush competition.

When people's mental models let them down in science, the result is usually just a lost opportunity. But as we have seen, when it happens in other spheres of activity, such as politics or war, the results can be disastrous. The failure of mental models also underlies many famous disasters.

## The Unsinkable Titanic

One of the most famous tragedies of the Nineteenth Century was the sinking of the Titanic, the supposedly "unsinkable" ship, on 14 April 1912. The accident claimed the lives of some 1,500 passengers and crew.[9] This famous tragedy provides a good example of the limitations of human foresight.

A large part of the mystique surrounding the Titanic disaster centres on the belief that the ship was unsinkable. This myth appears to have contributed indirectly both to the disaster happening, and to its magnitude. Even the master of the

Titanic, Captain Edward Smith, himself appeared to believe that the Titanic was unsinkable.[10] In conversations with passengers he is reported as stating that even if the ship were to be cut crosswise in three places, those pieces would still remain afloat.

In 1910 the White Star Line released a publicity brochure about its new twin passenger liners, Olympic and Titanic, which were then under construction. The brochure concluded by stating that "… as far as it is possible to do so, these two wonderful vessels are designed to be unsinkable."

Thus was born a myth in the public mind that the Titanic was unsinkable. In truth, the claim was more than just propaganda. The ship's designers had incorporated novel features to make the ship nearly unsinkable. For a start, the ship had a double-bottomed hull. This space not only provided an extra seal against hull damage, but also contained some 44 water tanks. Besides providing water for the ship's boilers, the tanks also formed ballast that kept the ship stable. Moreover the hull was divided into 16 compartments separated by bulkheads with watertight doors. These were designed to seal off breached sections in an emergency and to prevent the entire hull from becoming flooded.

In 1911, the Olympic became the first of the twin ships to be launched. Its reputation for unsinkability was enhanced when it survived ramming by a naval vessel. So the myth surrounding the two ships was well established by the time Titanic began its maiden voyage in April 1912.

At 11:40 p.m. (ship's time) on Sunday 14 April 1912, the Titanic struck an iceberg. Any four of the ships compartments could be flooded and the ship would still float. However, in a glancing blow, the iceberg scraped along the hull, tearing open five of the bulkheads. The hull rapidly flooded with water and Titanic sank less than 2½ h later.

In hindsight, several factors combined to cause the disaster, and belief in the ship's unsinkability probably contributed. No one conceived of a scenario in which more than four compartments could be flooded at once.

The magnitude of the disaster was increased dramatically by a shortage of lifeboats. Although the ship had 20 lifeboats, which exceeded the required standard at the time, even this number was far too few for the number of passengers on board.

At the time of the disaster the White Star Line ships were in fierce competition with the Lusitania and Mauretania of the Cunard Line. The Titanic's captain was under pressure to set a new record time for the trans-Atlantic crossing. Even though there had been reports of icebergs in the area he pushed the ship to its maximum speed.

The intense publicity that followed the disaster led to major improvements in safety. The International Ice Patrol was established to locate and report on the danger of floating ice and ships were required to keep up a 24 h wireless watch. New regulations demanded that passenger ships carry enough lifeboats to hold everyone on board in case of disaster.

To conclude, in all the kinds of accidents that were known at the time, Titanic was all but unsinkable. The problem was that no one thought of the unthinkable: an accident that was unknown; an accident that found the only way to overcome all of the ship's defences against sinking.

---

## End Notes

[1]From *Hamlet*, Act I. Scene V by William Shakespeare (1604).

[2]This account is based on information from Churchill (1923) and Horne (1962).

[3]Howard Hughes developed a morbid fear of germs during childhood. He spent the last years of his life hidden away in the penthouse of a Las Vegas hotel.

[4]Horne (1962), p. 18.

[5]Churchill (1923).

[6]For a detailed account see Wolfe (1979).

[7]Tenner (1996).

[8]Also, new technologies create new scenarios that were never anticipated when laws were written. The introduction of digital media for copying, storage and communication have led to situations that copyright laws never anticipated.

[9]The account given here is based on information provided by the *Titanic Historical Society*. http://www.titanic1.org/articles/titanicpastandpresent1.asp Downloaded 12 July 2013.

[10]Behe (2002). *George Behe's Titanic Tidbits*.
  http://ourworld.compuserve.com/homepages/Carpathia/page2.htm.

# The Sting in the Tail

**7**

*Every political good carried to the extreme must be productive of evil.*[1]

**Abstract**

Society has evolved many ways to eliminate extreme situations and to reduce their impact. Institutions such as emergency services deal with extremes. Laws and customs seek to eliminate extreme behaviour. Standardization aims to remove isolated products and practices. Media reporting distorts public perception of issues by focussing on extreme events as though they are the norm. Many social problems stem from failure to recognize extremes for what they are. One is prejudice, such as the false assumption that men are better than women at chess and other games. Another is the refusal to recognize and respond to extreme danger, which has led to disasters such as the destruction of St Pierre in 1906.

## Biased Samples

Lapland is a cold country. The home of Santa Claus and his reindeers. A land of ice and snow. At least I always thought it was, until I visited Lapland on a business trip a few years ago. The funny thing was that my visit happened to coincide with mid-summer. At midnight the sun was still shining brightly. I was in the country for an entire week, but the sun never set. And it was not just any summer. It turned out to be the hottest summer on record. The temperature soared. People went swimming in the lakes. There was no air conditioning, so we opened doors and windows to let a breeze flow through and cool the offices. Based on my experience, it was difficult to visualize Lapland ever being in the depths of winter. Despite all

D. G. Green, *Of Ants and Men*, DOI: 10.1007/978-3-642-55230-4_7,
© Springer-Verlag Berlin Heidelberg 2014

**Fig. 7.1** An example of rare events occurring over time. Purely by chance, there are long periods with no events at all, and clusters of events concentrated in a short interval of time[2]

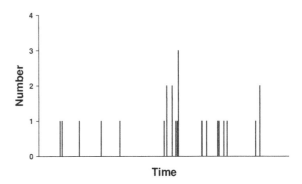

that I knew about harsh Arctic winters, I left with impressions of a country forever basking in the heat of mid-summer sun.

False generalizations of this kind are easy to make. It is dangerous to take an example, based on a personal experience close to hand, and assume that it is typical of the whole. It is all too easy to take the exception as being the rule; to confuse the unusual with the mundane, and to mistake an extreme case as being the mean (average).

Means and extremes arise in many social settings. Society has had to devise many ways to convert extremes into means. Confusion between the two is also a source of many social problems.

## Means and Extremes

Throughout history, no aspect of military tradition was ever more glorious than the cavalry. It is no wonder then, that leaders of the Prussian Cavalry grew alarmed during the 1890s when a bizarre series of fatalities threatened to mar their reputation. Suddenly, it seemed that horses were going crazy and kicking soldiers to death. Worried that something odd was affecting the horses, the Prussian High Command called in a noted mathematician, Professor Ladislaus von Bortkiewicz, to investigate. After examining data from 14 cavalry regiments over a 20 year period, von Bortkiewicz was able to show that a sudden increase in the number of deaths was not a sign of something odd affecting the horses. Instead it was just the extreme tail end in a natural distribution of rare events.[2] Low numbers of accidents, usually zero, were the norm, but years with high numbers of accidents, even a series of years, do sometimes occur (see Fig. 7.1).

Many events in society occur in the same kind of way. Car accidents are rare events in any given household. But sometimes a spate of accidents does occur. The important thing to understand about rare events is that when you enlarge the scale, the events become increasingly predictable. At any single intersection on any single day, car accidents are very rare. But across an entire city, and over the course of an entire year, the numbers become very regular. In the same way, the number of diners entering a café during any 1 min period is extremely variable, but over the course of an entire day, the numbers become highly predictable.

Road accidents and diners entering restaurants are just two examples of events that vary in random ways. Many other things also vary randomly. Take height, for instance. According to the *Guinness Book of Records*,[3] the tallest person who ever lived was Robert Wadlow. He was 272 cm (8 ft 11.1″) tall when he died in 1940, aged just 22. The world's shortest person was He Pingping, who grew to a mere 74.6 cm (2 ft 5.37″) in height. These people are the extremes. Very few of us are anywhere near so tall or so short. Everyone else lies in between. Most of us are close to average height. This average varies from country to country, but is around 173 cm (5 ft 8″) for men and around 160 cm (5 ft 3″) for women.[4]

Science is very bad at dealing with extremes. Traditional science relies on repeated testing by independent observers. But independent observation is difficult if the phenomenon concerned is too rare to be observed easily. Ball lightning, for example, was a phenomenon described many times by witnesses, but for a long time physicists doubted its existence because they could not explain it, nor observe it at will.

Remote regions of the Earth have long harboured animals and plants unknown to science. Tales of the gorilla, the okapi, and orang-utan were dismissed as native legends until scientists were actually able to observe them. The coelacanth, a fish thought to have been extinct for 200 million years, shocked the scientific world when fishermen brought a specimen back to shore in 1938.

From the Dodo to the Tasmanian Tiger, humans have wiped out many species. Even the Passenger Pigeon, once prolific in North America, fell victim to the spread of civilization. The problem is to distinguish extremely rare species from extinct ones. What no one knows is how many unknown animals have gone extinct as the spread of human disturbance outpaced the ability of science to fill the gaps in our knowledge. What no one knows is how many rare species, known or unknown, are presently teetering on the brink of extinction.

## All Roads Lead to Disaster

Fred and Daisy live on a farm a few miles down a country road from the nearest town. Every day Fred drives into town to pick up supplies and deliver goods to market. He loves this trip. He knows every inch of the road, which is usually empty of traffic. He can drive as fast as he likes and make believe he's a grand prix racer. Then one day, unknown to Fred, a truck hits a bump, which tips over a drum full of oil it is carrying. Oil splashes out and spreads all over the road. Next day, Fred charges around a sharp corner in the usual way and hits the patch of oil. The car begins to skid. Fred's surprise turns to alarm when he realizes that neither the brakes nor the steering have any effect. He has lost all control. The car slides across the road and slams sideways into a tree on the far side.

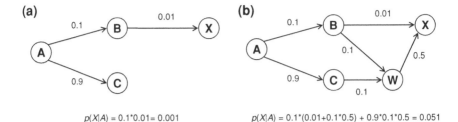

$$p(X|A) = 0.1*0.01 = 0.001 \qquad p(X|A) = 0.1*(0.01+0.1*0.5) + 0.9*0.1*0.5 = 0.051$$

**Fig. 7.2** A simple example showing how a richer network of events can increase the probability of a particular outcome (*X*). The *letters* and *arrows* denote sequences of events, and the *numbers* are the probabilities of each pathway occurring. In case (**a**), there is only one path from *A* to *X*, so the probability of *X* is 0.001. In case (**b**), the presence of an extra event (*W*) creates new pathways to *X*, so the probability of *X* rises to 0.051

Patterns of connection in networks can help us to understand how rare events can become inevitable. In Chap. 3, we saw that a game of chess is a sequence of moves that take one arrangement of pieces to another. In a similar fashion we can think of each day as a sequence of individual events, and this sequence is a pathway through a network of all events that could possibly happen.

Take a look at Fred's accident again. We can regard this as a sequence of events. For simplicity we could represent "Fred drives into town" as event A. Event X might be "Fred's car crashes". We can then look at pathways within the network of all possible events leading from A to X (Fig. 7.2).

Suppose for instance, that event A can lead on to several other events, say B or C. Of these, only event B (which might be "Fred takes the corner too sharply") leads to X. Event C might be the alternative ("Fred takes the corner cleanly"). If there is just one such path to X, then the chances of getting from A to X are not great (Fig. 7.2). But suppose that something changes so that another event, say W, can occur and this also leads to X. In the story above, event W might be "the car strikes oil on the road." It might also be other events such as "it rains", "Fred becomes tired", or "Fred gets distracted." Then the chances of getting from A to X increase considerably, especially if there are multiple pathways between all these events (see Fig. 7.2).

An important field of research, called Bayesian Networks, has grown up to tackle the challenge of analyzing causal pathways.[5] Its main uses are to support decision-making where many complex issues are involved. It has a wide variety of applications. Examples include medical diagnosis and treatment, environmental management, architectural design and pattern recognition.

## Managing Social Extremes

Extremes are a curse for society. Society functions best if it is in equilibrium. Riots, storms, floods, plagues and other disasters, big or small, disrupt the normal social routine and can lead to widespread misery and suffering. So it is essential that communities deal with extreme events. Society has therefore evolved many social institutions that serve to minimize the impact of extremes and keep life on an even keel. So one of the key functions of civilization is to convert personal extremes into community means.

Some of the most obvious examples of this can be seen in civil infrastructure. Take water supply, for example. People need water on a daily basis. Ultimately modern cities rely on rain to deliver the water they need. However rainfall is not a constant. It is intermittent, unpredictable. In dry regions rain may literally be a rare (that is "extreme") event. By trapping water in dams, we convert "extreme" rainfall events into a constant supply that people regard as the norm.

Many social institutions convert extremes into means. Building a house is an expensive affair. So too is starting a new business. Creating anything new requires an extremely large amount of money in a very short time. In most cases, the cost is far greater than people or organizations have ready to hand. Banks convert the need for extremely large amounts of money in sudden bursts into regular, manageable payments over a long period. The mirror image of a bank loan is insurance. You pay an insurance company small amounts of money at regular intervals just in case you need a large amount all of sudden.

Fortunately most of us never suffer the ordeal of a house fire. Major fires are a rare occurrence so most of us have neither the experience, nor the equipment to deal with one. The fire brigade has both of these things, and deals with major fires on a daily basis.

One man's extreme is another man's mean. Emergency services convert the problem of coping with events that are rare and extreme into an everyday business. Robberies, assaults, and other personal crimes are hopefully rare events in a person's life and the victims are ill equipped to pursue villains. But for police, tracking down and apprehending criminals is all in a day's work. Similarly, medical emergencies are rare in any household or workplace, but they are an everyday business for doctors and paramedics.

The business of converting personal extremes into social means extends well beyond emergency services. In reality it is the basis for all forms of specialization and the consequent division of labour.

A few years ago a friend completed work on his new house. He built it himself. The job took him several years and the work absorbed all his spare time. The biggest challenge, and one he greatly enjoyed, was learning how to do each part of the job. He was already a skilled carpenter, but he had to learn about plumbing, electrical wiring and many other issues virtually from scratch.

Few of us have the time and inclination to devote all our energies to building a house ourselves. There is simply too much else to do. Instead we pass the job over to builders, specialists for whom building houses is an everyday business. The same principle applies to almost everything else in our daily lives. Some people enjoy being home mechanics or building their own sound systems, but most of us turn those jobs over to specialists.

Commerce is founded on such specializations. Instead of trying to be a jack-of-all-trades, specialize in the one trade at which you excel. You can then exchange your skill for the specialized skills of others. Economists have suggested that specialization explains the emergence of markets. Adam Smith, for instance, suggested that:

> By pursuing his own interest he frequently promotes that of the society more effectually than when he really intends to promote it.[6]

Other economists, notably Hayek, have argued that specializations lead to spontaneous self-organization within society.[7]

Converting extremes into means is an important issue in business. In a small company, individuals may need to perform many different kinds of tasks. Most of the time a plumber mends pipes, but also has to spend time paying bills and completing paperwork. Larger companies achieve efficiencies of scale by hiring a specialist to handle the paperwork. This allows the rest of the workers to focus on their own specialities.

Society's practice of eliminating extremes also extends to the ways in which people think. In primitive societies, children learn from their parents. They absorb essential lessons about (say) hunting and gathering and about their traditions in the course of everyday activities. In modern Western culture, however, there is a mountain of technical skills that children need to acquire: reading, writing and arithmetic, and these are only the start. To ensure that children acquire all the standard knowledge and skills required by modern civilization, the practice of teaching has become a specialization.

One of the most important lessons children learn is to conform. Traditional education in most countries is a process of indoctrination and conditioning. Children are taught how to think and how to behave. Over a period of 12 years or more, children are conditioned to think and act in a standardized way. They learn to sit, calm and placid, for hours on end. They learn to follow every instruction given to them by their teachers. They learn to think and act within socially accepted norms and to avoid questioning the order of things. In short, traditional education is designed to eliminate wild, extreme ways of thinking and behaving.

There are many other ways in which society smoothes out extremes into means. Although not so obvious as those we looked at above, some are just as important. Social beliefs and values serve as models that define norms of behaviour. Reinforced by peer pressure, they discourage odd or extreme behaviour. For more serious issues, social norms are inscribed in law and enforced by police. Ideally these measures eliminate extreme kinds of behaviour, such as robbery and violence, by setting strict, formal bounds on the kinds of behaviour that are acceptable.

Surprisingly, in at least two areas—the law and politics—society actually encourages extremes. The aim in both cases is to find a balance between opposing views. In practice, it tends to polarize issues, which can distort outcomes.

For the most part the legal system is about eliminating extremes that are detrimental to society. But the actual practice of the law hinges around extremes. Cases brought before a civil court have two sides: a plaintiff (or claimant), and a respondent (or defendant). Each side argues for one extreme or the other and the judge or jury weigh the arguments on each side.

In a criminal trial, both sides try to pick holes in the case presented by the other. A friend of mine told me that he gave up criminal law because it was too easy to get guilty defendants acquitted. He gave the example of a drug addict who stole some bottles of tablets from a pharmacy. The thief was caught red-handed as he left the store. But when the case went to trial, he was acquitted because the prosecution could not show that the bottles he had with him were the same ones missing from the shelves in the store.

In civil cases, the fairest and best solution is often a compromise. But where lawyers' incomes depend on the result (especially those touting "no win, no fee"), their priority is to win, not compromise. The result can mean ruinous legal fees for the loser.

The polarization has several social side effects. For one thing, it effectively denies justice to most individuals: they simply cannot risk the prospect of ruin if their case fails. Another side effect is to spread the costs of litigation across society. Suits over damages have led to tighter rules around negligence and "duty of care".

Excessive litigation has led to a ludicrous situation where some individuals take no responsibility for their actions and where organizations must try to anticipate every kind of accident that stupid behaviour might lead to. One result is notices about obvious safety precautions. For example, buildings display signs instructing people to hold on to the handrail while climbing stairs. Packing boxes have labels warning "lift with care." Coffee cups have labels saying "caution: hot." The implied fear is that the building owners will be held responsible for any idiot who does not take this obvious precaution. Similar concerns have contributed to the "nanny state", which we look at in Chap. 9.

Another side effect is that insurance fees spiral to meets the costs of litigation. In several countries, medical malpractice suits have led to massive increases in insurance premiums, which doctors have to pass on their patients as increased fees. Likewise in my own home town, litigation has led to huge increases in public liability insurance. This has impoverished society in unexpected ways. In one case, a district football competition for children had to be cancelled because the league could no longer afford the insurance premiums.

The other arena in which society encourages extremes is the political system. In the Westminster system of parliament, there is a government and an opposition. The government proposes new courses of action and, almost without exception, the opposition opposes them. This creates a serious problem for any initiative that society needs. Whatever it is, the opposition will not support it. Instead, they will pick at it to find any flaw they can exploit for political advantage. So instead of society achieving consensus and support, any action will leave some people disgruntled.

This problem makes governments extremely cautious about the actions they take. They know that if their actions annoy too many people, their popularity will wane and they will lose power at the next election. This problem is especially acute for actions that achieve long-term benefits at the expense of short-term pain. This risk makes governments reluctant to take any long-term action. It also leads them to take actions that are popular, rather than addressing the real problems. In recent times, with opinion polls held almost daily, it leads to government actions being dictated by the latest polls, rather than by sound planning based on evidence.[8]

## Eliminating Extremes

Industrial standards, such as battery size, reduce the complexity of the supply networks that provide essential products (see Chap. 3). In effect they eliminate extremes (odd designs) and link everything to means (standard designs).

To most people, standards probably rank among the most boring topics imaginable. If you want to kill a conversation, just start talking about standards. Even if they have heard the term, most people would be hard pressed to name any standards. At school many of us learned about a metal rod stored in a special chamber in Paris and used as a standard metre. But what other standards are there?

It would surprise most people to learn just how many standards there are. Units of measurement are just the beginning. If you want to buy a new shoe, then the search is speeded up enormously because of standards for shoe size. If you need to replace a light bulb, you can thank standards for ensuring that the bulbs will match the size and nature of the socket. When you fill your car with petrol, you rely on standards to ensure that the hose will fit. You can thank standards for simplifying the supply of thousands of anonymous parts that go to make up your car and your appliances. You can thank standards for ensuring that your home does not fall down and that you do not get food poisoning every time you buy take-away food.

In general, standards are systems for ensuring widespread uniformity in some activity. They are essential in any area of activity where people have to bring together elements from different sources. For instance, a home computer is built from components that all have to fit together. The network card needs to fit into a compatible slot. The video card needs to output signals that the monitor can display.

The International Standards Organization (ISO) is the world umbrella for defining and coordinating standards in over 130 countries.[9] It lists thousands of different standards, some of them highly technical. Each year about 15,000 experts contribute to the development of new standards and the refinement of existing ones.

It is no surprise then, that standards also form a crucial element in computing. Many computing standards are well known. In a very real sense, the Internet itself is nothing more than a set of standards. For example, the Internet Protocol (IP) is a standard that defines a consistent way of addressing locations across different networks. The World Wide Web is built on the HyperText Transfer Protocol (HTTP). This defines what happens when you click on a link in a Web page.

Standardization is one of the more subtle ways in which information technology contributes to globalization. Large companies rely on economies of scale to give them an edge over the competition. However, efficiency and cost effectiveness are difficult to achieve if operations are not standardized.

Suppose, for example, that a company manufactures a product that relies on widgets from local suppliers. But if these widgets are made in (say) imperial sizes in the US and metric in other countries, then the local widgets may not be useable. Or again, if each division of a company sets up its database using different software, then sharing information and compiling centralized records becomes a difficult and costly affair. So standardization is absolutely crucial to the viability of any kind of advanced industry.

Now, suppose that a lot of different organizations operate to identical standards. If two companies produce the same item, the one that can sell it cheaper is likely to drive out the other.

Standards do have other side effects. One is that once locked into a particular standard, it can be difficult to change. While most of the world uses metric standards (metres, kilograms etc.), the USA persists with old imperial units (feet pounds etc.). In 2009, NASA estimated that to convert all its documentation to metric units would cost around $370 million.[10]

## From Bias to Prejudice

As we saw earlier, biased samples can lead to false impressions. Even a single bad experience may colour a person's opinions forever. A child is bitten by a dog and fears all dogs forever after. One rude remark from a shop attendant can be enough to drive a customer away permanently. A girl has a bad experience with a boy she is dating and decides that all men are monsters. There are many ways in which biased samples can draw us unwittingly into prejudice.

Air crashes are rare. They are extreme events. Statistically, travelling by air is safer than driving.[11] The trouble is that air crashes usually kill many people at once. They make headline news. One result is that some people who think nothing of driving on busy city streets develop phobias about flying.

The reporting of air disasters is just one way by which media tend to create an impression that rare or extreme events are the norm. Newspapers and TV will report any sensational event in lurid detail. Major crimes are a rarity in most neighbourhoods, but they are presented as daily events in media and in newspapers.

It is a principle of reporting to avoid bias by presenting a "balanced" view. In practice, presenting a balanced view usually means presenting both sides of a dispute. But if one side of the dispute is (say) an extremist minority, then providing both sides with equal air time is actually promoting the extremist view. So in their attempts to avoid bias, reporters sometimes succeed only in promoting it.

Unfortunately the media do not extend the principle of a balanced view to the selection of news stories. News everywhere is biased towards sensational extremes, especially violent crimes, accidents and scandals. Movies likewise portray extreme situations as though they are commonplace.

Other aspects of the media also portray extremes as means. Promotions for lotteries always convey the impression that if you enter, you are bound to win first prize. The focus on image has a similar effect. The glamour surrounding the top movie stars draws in thousands of young hopefuls, convinced that they too will become rich and famous. What you do not normally see are the thousands of excellent performers who never get that big break, and struggle to find any work at all. This truth is evident in popular reality shows for would-be performers. They attract thousands of hopefuls, many of whom are excellent performers, but they are rapidly cast aside in a search for the one eventual winner.

Confusing the exceptional with the ordinary leads to many kinds of problems. A marketing company might look at its most prolific salesperson and make her results the standard that everyone is expected to achieve. In a cost cutting exercise, a manager reduces monthly spending by firing staff, and then gets told he has to make further percentage reductions of the same size every month.

Failure to recognize changes in a distribution can also lead to false impressions that create prejudice. In the book *Full House*[12], Stephen Jay Gould asked the question: "Why are there no 400 hitters in baseball today?" In the early 1900s, some baseball players were able to achieve sensational averages, but not today. The same effect is seen in many other sports. In cricket, for instance, Don Bradman had a lifetime batting average of 99.96 during the 1930s and 1940s, but the best modern batsmen cannot manage an average more than 60.

Does this mean that players today are not a patch on those heroes of a century ago? No. The real problem, as Gould pointed out, is that today's players are better! By this he did not mean that the best players are better than the best of years ago, but that the *average* player today is better than the average player of 100 years ago. The average pitcher is now better than 100 years ago, so for batters, it is as if they were always facing the best pitchers.

The failure to understand the nature of distributions can be used to prop up prejudice. For instance, at the time of writing, there are only three women amongst the top twenty ranked chess players in the world. Is this proof that women are no good at playing chess? Does it show that men are smarter than women? Not at all. A careful study of the issue showed that:

> ...the great discrepancy in the top performance of male and female chess players can be largely attributed to a simple statistical fact—more extreme values are found in larger populations.[13]

On average women perform just as well as men at chess. It is a difference in the numbers of players that tells at the extremes. More men play chess, so there are more men who are extremely good at chess (Fig. 7.3).

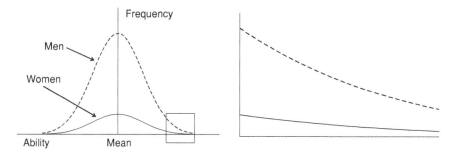

**Fig. 7.3** Confusing extremes and means in chess. The graph on the *left* shows the distributions of skill for women *solid line* and for men *dashed line*. The *horizontal axis* represents increasing skill. The *vertical axis* represents the number of players. The average level of skill (marked by the *vertical line*) is the same for both men and women. The section in the box (the frequency of excellent players) is expanded on the *right* and shows why it is that more men appear in the top rankings than women

## Force of Habit

Force of habit can be a dangerous thing. For people who drive to work, for instance, and take the same route every day, the trip can become so routine that they tune out. They daydream, they talk to their passengers or they listen to the radio. The lowered attention to the task can cause them to respond more slowly than normal to dangerous situations. As we saw earlier in this chapter, the result will be to increase the chances of a dangerous event occurring. So if people have bad driving habits, such as following too closely then there is a real possibility of an accident.

Maintaining concentration at any repetitive task is difficult. If your concentration flags, then you are likely to react instinctively and may fail to notice changed or unusual circumstances until it is too late. This problem is a widespread source of accidents, including industrial and domestic accidents, as well as on the roads.

The problem with the way we learn is that our understanding of the world is limited by our experience. Instead of general principles, we learn simple recipes that are based on individual experiences. This is not to say that we do not generalize at all. No, we generalize incrementally, by assimilating one similar experience after another. To take the earlier example of entering a room, our mental model of a room emerges out of many encounters with different rooms.

We have a limited knowledge of people and events and this makes it impossible to recognize all the possibilities. So we tend to assume that everything fits into the scheme of what we know. This opens the way for the platypus effect (Chap. 6): our learning fails us when events happen that are outside our experience. Fortunately most events fall within the mean—the typical, the predictable. So, just like George waiting in the bank queue (Chap. 4), it is the extremes that have the greatest impact as experiences.

The same happens to entire communities. If society has many mechanisms to convert extremes to means, then situations that it fails to convert are the cases that will disrupt it utterly. Nicholas Taleb argued that improbable (extremely unlikely) events play an important part in shaping social trends:

> Almost everything in social life is produced by rare but consequential shocks and jumps; all the while almost everything studied about social life focuses on the "normal," particularly with "bell curve" methods of inference that tell you close to nothing.[14]

He called such a consequential event a *Black Swan*, because like the black swan of Western Australia, it exhibits three attributes:
- It is rare; an outlier to normal expectations and experience;
- It has extreme impact;
- It has retrospective plausibility.

According to Taleb, the 9/11 terrorist attacks on New York and the success of Internet giant Google were events of this kind. Both events had widespread impact. Neither was predicted beforehand, but seemed inevitable after the fact.

## A Frog in Hot Water

A well-known story goes like this. Place a frog in hot water and it will jump out immediately. On the other hand, if you place a frog in cold water and gradually heat the water, then the frog will sit there until the water boils.[15] Like a frog in hot water, people sometimes become so adjusted to deteriorating circumstances that they do not seek to escape, even under conditions that they would otherwise find intolerable. Sometimes it arises because victims cannot see any way out. Sometimes it occurs because people prefer the familiar, no matter how bad, to the unknown. And sometimes, it occurs because people want to cling on to what they had so much, they simply deny reality.

A danger in the confusion between means and extremes is that people become too accustomed to whatever seems normal for them. Being accustomed to their everyday life, some people no longer accept the possibility that extremes do exist. One of the most striking aspects of the forests near where I live is their unchanging appearance. Year after year it appears as if nothing ever changes, as if the forest has been this way forever. And yet reaching way above the tall trees there are giant black and white skeletons of old trees. When I was young this was a mystery to me, but it turns out that these skeletons are the trunks of tall trees that were burnt out in disastrous fires that devastated the area years before I was born.

The assumption that there are no extremes—that things never deviate from the mean—is also dangerous. In our day-to-day lives, it is easy to assume that things will never change. Children assume that their parents will always be there for them. We assume that power, water and other public services will always be available. And yet, just a few weeks ago, heavy storms brought down power lines everywhere in my home town, cutting out power to half the city. Traffic came to a standstill and some areas were without power for over 24 h.

When rulers and public officials assume that nothing will change, the results can be catastrophic. A dramatic example of this happened on the Caribbean island of Martinique in 1902.[16] On April 25, 1902, Mount Pelée began erupting. Over the next two weeks the eruptions continued, gradually building in intensity. In St Pierre, Martinique's capital city, 4 miles to the south of the mountain, citizens were subjected to earthquakes, falling ash and the unpleasant smell of sulphur in the air. As wildlife tried to get away from the volcano, the city had to contend with plagues of ants, centipedes and poisonous snakes.

Despite the increasing problems, newspapers claimed the city was safe. As we saw in Chap. 5, denialists may refuse to accept what is obvious but unpalatable to them. In St Pierre, the authorities refused to accept the need to evacuate and even refused clearance for ships to leave the harbour. In doing so, they condemned the entire population. When the main eruption occurred, the city was still full of people. On May 8, 1902, a pyroclastic cloud swept over the city, instantly igniting everything in its path. St Pierre was destroyed and 30,000 people were killed.

The confusion between means and extremes impinges on some of the most contentious issues of our time. One of the most hotly contested is the question of climate change. On a technical level, this question boils down to a debate about means versus extremes. On the one hand, advocates point to the extremes in weather that have been observed over a period years. On the other hand, skeptics argue that what we are seeing as "extremes" in recent weather records are actually well within the range of climatic variations that have occurred over the past few thousand years. However, the really crucial question is whether current climate trends are natural or side effects of human activity. We will come back to this issue again later (see Chap. 19).

## End Notes

[1] Attributed to Mary Shelley.
[2] This is known as the Poisson distribution.
[3] Guinness world records (2010). www.guinnessworldrecords.com/records/human_body/extreme_bodies/.
[4] The average height varies around the figures given here, being about 5 cm more. See Wikipedia (2010). *Human height*. http://en.wikipedia.org/wiki/Human_height
[5] For a detailed technical introduction to Bayesian Networks see Korb and Nicholson (2011).
[6] Adam Smith (1776).
[7] For example Hayek (1982).
[8] Megaloganis (2010).
[9] http://www.iso.org/iso/

[10]Marks (2009).

[11]According to the US National Safety Council, in 2009 the rate of fatalities for "light duty vehicles" was 0.53 deaths per 100 million passenger-miles, while the same rate for air travel was a mere 0.01. Source: NSC (2012). *Injury Facts*. p. 154. http://www.nsc.org/news_resources/Resources/res_stats_services/Pages/FrequentlyAskedQuestions.aspx.

[12]Gould (1997).

[13]Bilalić et al. (2009).

[14]Taleb (2007).

[15]The legend of the boiling frog is an analogy. I have not tried this experiment and do *not* recommend that anyone try it. For a wider account of its use in publications see http://en.wikipedia.org/wiki/Boiling_frog.

[16]This account is based on Thomas and Witts (2011).

# Divide and Rule

**8**

> *We can form a single united body, while the enemy must split up into fractions. Hence there will be a whole pitted against separate parts of a whole, which means that we shall be many to the enemy's few.*[1]

**Abstract**

The usual way of dealing with complexity is to try to reduce it by "divide-and-rule". You carve a big problem into smaller, self-contained parts that are easier to deal with. Familiar examples range from designing houses with bedrooms and bathrooms, to dividing large organizations into sections with specialist roles. This classic approach leads to social hierarchies as large issues are divided repeatedly into smaller and smaller roles. It also results in unexpected problems when events cut across several parts at once. Compartmentalization and classification promote limited, closed-box thinking. One example is the tendency to reduce everything to simplest possible terms, especially numbers, which hides their true complexity.

## Why is a House Like a Handbag?

Has something like this ever happened to you? You are in a hurry to get to work, so you throw all your effects into your handbag and off you go. But then your phone rings. Buried deep inside your handbag, it is hidden by dozens of other items. While you rummage around searching for it, the phone rings once, rings twice, three times. At the exact moment you find it, the ringing stops.

Handbag manufacturers are aware of this problem. A cheap handbag is just a bag. But better ones are well designed. They provide pockets and compartments to help you organize your belongings and reach them easily. Pockets provide control

D. G. Green, *Of Ants and Men*, DOI: 10.1007/978-3-642-55230-4_8,
© Springer-Verlag Berlin Heidelberg 2014

over your possessions. Pockets enable you to store your phone, your wallet, your keys so you know exactly where they are, and where you can reach them when you need them in a hurry.

In effect pockets simplify the problem of storage and access by breaking it down into smaller parts. Problems occur if you neglect to put things back in the right place. Throw your key into your bag, not its pocket, because you are in a hurry and it will be "lost" next time you need it.

The essence of control is to ensure that a system conforms to a desired model. Take the example of a home. Like a handbag, a house contains many objects. But houses are designed around the way people live. So it is obvious how to organize things: pots go in the kitchen; towels go in the bathroom, clothes in the bedroom, and so on. Ideally there is a place for everything and everything is in its place.

Even if it has places assigned for everything, a house will quickly get in a mess unless people put things away every time they use them. This becomes difficult if the home owners are busy people. It becomes well nigh impossible if they have small children.

Even if people are absolutely fanatical about putting things away, their system can break down if they keep acquiring possessions. When I moved interstate years ago, all my possessions fitted comfortably into a suitcase and a single briefcase. To move house twenty years later, we needed two large trucks and a car towing a trailer. Unless you throw out as many goods as you acquire, the volume of possessions inevitably increases over time. As the volume approaches the maximum available storage space, finding places for everything, and keeping them there, is an increasing challenge.

## Divide and Rule

Sally is frantic. She's engaged to be married to Frank and the wedding is less than a month away. It's to be a big wedding. The bride and groom each have big families and loads of friends. So there are heaps of things to do. Sally's problem is that she has been trying to organize it all by herself. Overwhelmed by the complexity of it all, she cannot cope. Things seem to be falling apart. In desperation she rings her best friend and bridesmaid Suzy.

"Well" Suzy sighs, "If you'd accepted our offer to help a few weeks ago you'd have it all done by now." An hour later Suzy and the other bridesmaids turn up at Sally's house. They split up the work between them. Suzy organizes the invitations, and the others deal with the reception, catering and so on. All Sally has to do is answer questions about her preferences. Feeling back in control, she's much calmer now. The wedding goes off without a hitch.

Sally and her friends have used the age-old principle of divide and rule to deal with her wedding plans. This is the traditional way of coping when things grow complex. And it is very effective. When faced with a large, complex problem people cope by breaking it down into smaller, simpler, and more manageable problems. In effect, you deal with the complexity by eliminating it. This approach applies almost everywhere, from managing organizations to solving problems, to organising our homes.

We reduce complexity by carving bundling things into separate modules. This is done by grouping related parts of a system and isolating the group from the rest of the system (see Fig. 8.1). To run a large company, for instance, managers usually dividing its workload into logical units that carry out well-defined and separate roles. These units are usually functional, such as divisions of finance, manufacturing, marketing, or they may be geographic, such as plants and regional offices.

To understand the need to simplify, recall the Travelling Salesman Problem from Chap. 3. No salesman has the time to compute billions of possibilities to find the very shortest route possible. What a salesman does in practice is to break the overall problem down into smaller, simpler problems. For instance, he might make a circuit of several major cities and, based in each one in turn, make circuits around the towns nearby. This solution may not be the very shortest possible route, but it is simple to figure out and adequate for their purposes.

In engineering the divide and rule approach appears as modular design and construction. Manufacturers make modular components. Modules simplify a system by reducing the number of interactions between the parts. Thousands of parts go into the making of a motor car, for instance, but these are arranged into functional modules such as the engine block and the dashboard control chassis. Modular construction makes it easier to manufacture and assemble cars, as well as to service and repair them.

The idea of modular design has also evolved in nature. Trees, for instance, grow in modular fashion. They simply add the same modules again and again and again: branches, leaves, buds and fruit. The human body is also modular, with distinct organs such as the heart, lungs, liver and kidneys each carrying out vital functions. The bodies of insects and other arthropods have a modular design that consists of a head followed by a sequence numerous body segments, each containing a pair of legs. Even the genome is modular. Groups of genes switch on in coordinated fashion during development. Often, as in the case of the eye, for instance, the operation of a group of genes is switched on and off by a single controller gene.

The idea of modularity is deeply embedded in our culture, and in our thinking. We have social institutions, especially the government and the legal system, to control and manage social complexity. Small tribes and communities do not need laws. They are able maintain social discipline by peer group pressure.[2] But in large, modern cities, most people are strangers to one another. Rulers introduced laws, for example, to eliminate undesirable interactions within large groups of people. Unfortunately, those laws are never perfect. Human behaviour and the ways people interact with one another are so complex that it is impossible to codify them completely. Lawmakers can never anticipate every permutation of behaviour and circumstance. The legal system constantly faces the challenge of trying to plug holes that were missed by the lawmakers. The arrival of the Internet, for instance, suddenly revealed enormous holes in copyright laws. The advent of

**(a)**                                          **(b)**

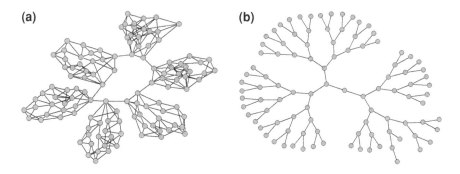

**Fig. 8.1** Examples of modular networks. **a** A network with few connections between modules, but each module well connected internally. **b** A tree, with a root node in the centre. Each branch provides the link into a module

cloning and genetic engineering challenge many long-held assumptions about patent law, not to mention morality.

From antiquity to the present day, rulers have used the principle of divide and rule to subdue subject peoples: keep rival tribes fighting amongst each other and they will have neither time nor inclination to fight you. Julius Caesar knew this, so did Machiavelli. The British applied this principle in many conquered lands, first in Scotland, and later in India, Africa and North America.

In the early 1800s, Port Arthur on Tasmania's southeast coast was one of Britain's most notorious penal colonies. Perhaps the cruellest aspect of the way they treated prisoners was to isolate them completely from other prisoners. The design of the chapel at Port Arthur is both bizarre and symbolic of cruelty. It consists of booths where convicts were seated during services. The booths were designed so that prisoners could see and listen to the service. But, they could neither see, nor communicate with any other convict. By isolating every prisoner from his fellows in this way, their captors could ensure that they had no opportunity to plan an escape. Isolate a prisoner from the support and encouragement of his fellows and you have a good chance of crushing his spirit.

We see the principle of divide and rule in many practices in modern society. Earlier (Chap. 7), we saw that specialisations simplify the way that society deals with the "extremes" involved in dealing with particular problems. Take the problem of building a house. In a few cases, friends of mine have built entire homes for themselves. However this is a very complex and time-consuming activity. Most people I know have engaged a builder to oversee the building of their homes. This role of this "builder" is to manage the work of contractors who carry out all the specialized tasks involved, including brick-laying, carpentry, plastering, plumbing and electricity installation. In other words, the "builder" breaks down the complex task of building and coordinates all the component tasks.

## Morning Tea at the Office

We see examples of the platypus effect (see Chap. 6) in the ways in which people interact with one another. Building design is a good example. Over the years, I have found myself working in many different offices and buildings. As architects are well aware, the design of an office has a large bearing on social interaction. In some buildings where I have worked, the offices are lined up next to each other in rows along a long corridor. Everyone was hidden away behind closed doors. When you walked along the corridor, you would rarely see your colleagues. There was very little interaction between people; they usually met only at set meetings. Many people knew little about the work done by their colleagues. Sometimes, they hardly even knew the person in the very next office. By way of contrast, another office where I once worked adopted a completely open plan. Everyone worked at desks and benches within a large open area. As a result, discussion between colleagues was commonplace, much too commonplace. Because of the constant interruptions, it was difficult to get any work done. Even if people did not talk to you directly, you could not help but hear their conversations on the phone, or with another colleague. Between phone calls, visitors and talkative colleagues, the level of interruptions would explode to the point where it was difficult to get any work done at all. I can even recall occasions when the there were interruptions stacked up on interruptions, as many as four or more deep!

The above examples represent extremes in interaction and connectivity, between staff. In the first case, every staff member is compartmentalized, literally. This arrangement allows each person to work but eliminates potentially fruitful interactions. In the second, there is nearly total interaction, and the activity is dominated by interactions rather than individual productivity. The best arrangement is a compromise between these two designs. Each worker has a separate office, but instead of lining a sterile corridor, the offices open onto a communal work area, so that people pass through on their way in and out.

---

## Climbing the Pyramid

In a tree structure, communication between separated elements is confined to pathways up and down the tree. For information to travel from any node $A$ to another node $B$, it must move up the tree until it reaches a node that lies above both $A$ and $B$. It then passes down the tree to $B$. In a large system, a tree structure is an efficient way to ensure full connectivity. For a well-balanced binary tree the average path length between $N$ nodes is $\log_2 N$. For example, in a network of (say) 1,000 computers, direct wiring to connect each machine to every other machine would require 500,000 wires. In contrast, a binary tree could link all of the machines using just 999 wires, but data transfer between machines would require hops between up to 18 intermediate machines. However, by adding more than two links at each node, the maximum path required rapidly diminishes. Among other things, this ability of hierarchies to connect all nodes efficiently makes communication via the Internet feasible. You can guarantee that a message can get from one node to another in no more than a certain number of steps (Fig. 8.2).

In management, hierarchies arise from the dual desires to simplify and to control complex organisations. They simplify control by the divide and rule approach. A manager at any node need only be concerned with the nodes immediately above and below. The model also restricts communication between arbitrary nodes in the tree. This effect enhances control, but can also limit the passage of crucial information, thus inhibiting responsiveness, efficiency and innovation.

In the early days of the World Wide Web, managers of large corporations and institutions became alarmed about the impact that the new medium was having. As the founder of one of these early Web sites, I experienced the hostile reaction of senior managers first hand. For two years they attempted by one means after another to close my web site, despite the enormous international publicity it was creating for local researchers. I soon realized that the source of the hostility was their fear of losing control.

In engineering, hierarchies take the form of modules. The advantage of modularisation is that it reduces complexity. This simplifies management of a large system. Any large system, such as an aircraft or a factory, may consist of thousands or even millions of individual parts. A common source of system failure is undesirable side effects of internal interactions. The problem grows exponentially with the number of parts, and can be virtually impossible to anticipate.

The solution is to organise large systems into discrete subsystems (modules) and to limit the potential for interactions between the subsystems. This modularity not only reduces the potential for unplanned interactions, but also simplifies system development and maintenance. For instance, in a system of many parts, the risk that some part will fail is always high. To cope with this problem, engineers build redundancy into crucial systems. Modularity also makes it easier to trace faults. In some systems (e.g. computer hardware), entire modules can be replaced so that the system can continue to operate while the fault is traced and corrected. In computing, the idea of modules has led to object-oriented programming, in which each object encapsulates a particular function and can be reused in many different contexts.

An important effect of modularity is to reduce combinatorial complexity. This effect is useful in many kinds of problem solving. To take a simple example, if an urban transport system has (say) 100 stops, then it is essential that travellers should be able to look up the cost of travel between any combination of stops. If every trip has a different price (based on distance, say), then a complete table of fares would need to hold 5,050 entries (assuming that the distance from $A$ to $B$ is the same as from $B$ to $A$). This number makes any printed table unwieldy. On the other hand, if the stops are grouped into (say) five zones, with every stop within a zone considered equivalent, then the table would need at most 15 entries, which would make it both compact and easy to scan.

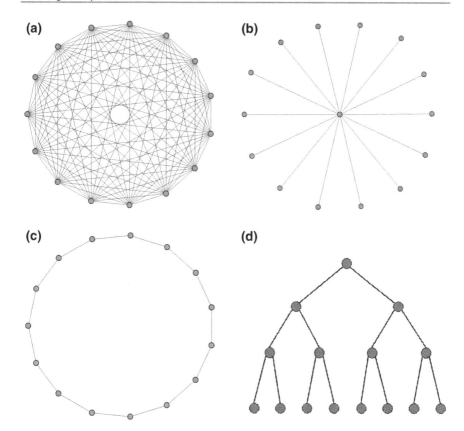

**Fig. 8.2** Designs for communication networks of 15 nodes. **a** Linking every pair of nodes provides efficient but requires 105 lines. **b** A star network limits separation to 2 steps at most, but puts a load on the central node. **c** A ring requires only 15 lines, but the average separation is 4 steps. **d** A binary tree is a good compromise, requiring just 14 lines, with an average separation of 3.9 (log2 15)

## The Brittleness of Modern Technology

Most of us like to get away from civilization from time to time. We like to get close to nature. I wrote part of this chapter on a laptop computer while sitting by a lake near my home. In such surroundings, it is easy to forget how dependent we are on the technology that makes our lifestyle possible.

If the survival of traditional cultures depends on having a critical mass of people to carry on the knowledge and traditions, then the same principle applies even more so to modern technology. Back in the days of the Apollo Moon Programme, many allusions were made to the effect that the astronauts walking on the moon were like the tip of a vast pyramid, with tens of thousands of engineers, technicians and other professionals supporting them. With a small army employed by NASA at the time, it was easy to see understand how complex the entire program was, even if the roles of those thousands who made up the pyramid was never spelt out in full. The same comparison holds true about all kinds of explorers, the mountain climbers who reach the summit are the representatives of what are sometimes very large support teams further down the mountain.

What most people never realize is that our entire technical culture is like that. Almost every product we buy, every piece of equipment we use, is the end result of contributions from literally thousands of people. Just look at the credits at the end of any movie. Many years ago, the credits never extended much past the cinematographer. Today everyone gets a mention and the credits can roll on and on listing hundreds of names. But even then, they do not fully list all the contributions. Many names are hidden under contracted organisations. But what about all the people who built the cameras, who manufactured the film stock, not to mention the caterers, drivers, hotel staff who looked after the cast and crew, and so on and so on.

Think of an example a bit closer to home. What does it take, for example, to make and run a computer? The chip is the most advanced component. It requires a specialized plant and highly skilled workers. But that's not all. There are many other components in a computer. The screen, the boards, the disk drives, power supply, the keys, wires, the chassis, the screws, and so on. These parts come from many different suppliers.

But there is more. The components in turn have to be manufactured. Each key on the keyboard, for instance, must be made in a plastic mould and then have letters painted on it. Each components of the motherboard has to be carefully manufactured and tested using precision instruments. These days, many of these steps are automated, but people still have to oversee them.

However the story does not stop there. Far from it. Like an onion, we can peel back the technology layer by layer, revealing more and more people who contribute to the process that put your computer together. The people who do the actual work of assembling the computer are really just the last stage in the process. Likewise, the people who build the components need to get materials, the metals, plastic, paint, and glass from other suppliers. And those suppliers in turn, get materials from other suppliers, such as the refineries that extract the metal and the mining companies that dig the ore from the ground.

All the way through the above processes, workers need machinery. And that machinery needs to be manufactured. And that manufacturing process is likely to be as complex as the process of building the computer itself. And to run that machinery, the workers need fuel and they need power. And the workers themselves need to be fed, and housed. And the food needs to be supplied from somewhere. And so it goes on, and on and on.

In short, building your computer is a horrendously complex process. The final machine emerges out of the joint efforts of thousands of people spread across hundreds of companies and other organizations.

The point is that maintaining any high technology requires a large, diverse workforce. It also demands the vast range of knowledge and skills that all those workers represent. Below a certain threshold, which is difficult to estimate precisely, many high-tech industries become impossible to maintain. In fact no country can really operate its high-tech industries all on its own. Whether it be the specialized parts, or rare materials used in manufacturing.

One way of appreciating the complexity of modern society is to look at the number of steps that separate end users of a finished item from the source of raw materials from which they are made. For most items in primitive societies, no more than three steps are required.

Step 1: The hunter catches food;
Step 2: His family fashions the hide into a coat;
Step 3: The hunter trades the coat for a new spear.

In modern society, the number of steps normally much higher, for instance:

Step 1: The farmer raises the sheep;
Step 2: Farm hands shear the sheep;
Step 3: The farmer delivers the wool to a wool corporation;
Step 4: A wool corporation sells the wool;
Step 5: The wool is turned into yarn in a factory;
Step 6: The yarn is knitted into a sweater;
Step 7: A wholesaler delivers the sweater to a retailer;
Step 8: The retailer puts the sweater up for sale;
Step 9: You buy the sweater.

The network also has many more branches. An automobile, for instance, is put together using hundreds of separate parts, leading to an ever-widening tree, such as we saw earlier.

## Thinking Inside a Closed Box

We see many examples of the divide and rule approach in the ways we think about complex problems. One is the tendency to treat each problem as a closed box. That is the situation is affects nothing outside and is in turn not affected by anything outside.

The mistake in local thinking is the assumption that our local, everyday activity is a closed box, unrelated to and not influencing anything that goes on elsewhere. We see this story in many guises. In each case, we see people doing drifting inside their comfort zone. They do what they know and ignore everything else. They ignore the bigger picture, the many consequences and implications their actions have on the wider world around them.

A few years ago an infamous photograph did the rounds of the Internet. It showed a dead porcupine lying in the middle of a road. Sad perhaps, but what makes it bizarre is that workers painting line markers have sprayed the new centre line right over the top of the corpse. Their job was to paint lines, not clean up road kill. Dead animals on the road? Not my problem.

There are many other examples of this kind. It leads to bureaucrats persecuting individuals regardless of extenuating circumstances or wider considerations. It leads bigots to blame a single group for all their problems. It encourages lawyers to place all the blame for an accident on a single person so they can sue them. Many of these cases are examples of categorization: we place people and things in categories: tall, short, good, bad, friend, foe based on simplifying criteria that fail to take into account all the complexities of human motivation and behaviour.

The problem with categorization is that it loses important details under a veneer of generalization. For instance, not all business people are anti-environment, not all poor people are criminals, and not all academics are boring intellectuals. Perhaps most telling of all is the categorization into local and remote: things that affect a person directly and things that do not, and can therefore be ignored ("not my problem"). This simplification leads directly to Merton's rule of the imperious immediacy of local interest.

## Playing by the Numbers

Treating complex things in simplistic ways is nowhere perhaps more evident than in our use of numbers. Numbers are everywhere in modern society. They dominate our lives, and our thinking. This is not to say that using numbers does not have its uses. Problems arise when numbers introduced for one reason come to be used for something else.

Properly used, metrics can reveal where serious problems lie. To take a familiar example, we can look at statistics in modern sport. As sport has become increasingly professional, the stakes get higher to achieve peak performance on the big occasions. So athletes try to iron out every aspect of their game. Advances in technology have helped to make metrics commonplace sources of evidence in assessing performance.

Take tennis for example. Here is a sample of the kinds of statistics that commentators at grand slams and other leading tournaments routinely provide about players:
* percentage of successful first serves;
* percentage of points won on first serve;
* percentage of points won on second serve;
* number of double faults;
* average speed of first serve;
* average speed of second serve;
* number of aces;
* number of winners;
* number of unforced errors;
* number of approaches made to net;
* number of points won at the net.

Glaringly high numbers anywhere in the table point to strengths or weaknesses in a player's game. A low percentage of first serves tells a coach that the player needs to work on his or her serve. The danger of such statistics is that players and coaches can get drawn in and focus too much on correcting the numbers at the expense of other aspects such as mental strength, tactical skill and the ability to spot an opponent's weaknesses.

The corporate world tends to reduce people to numbers. In a small organization, the manager knows everyone and sees how they perform on a day-to-day basis. But large corporations become more impersonal. So they measure people's performance using Key Performance Indicators (KPI). This approach reduces everything a person does to a set of numbers. The basic idea is to assess how an individual contributes towards the corporate goals. Usually the ultimate goal is to increase profits. Each indicator is meant to reflect some factor that promotes that goal. So the indicators may be number of sales per month, production output, or size of cost cuts.

The problem with summarizing complex activities and priorities as a set of numbers is that maximizing those numbers becomes the only goal for the individuals involved. And the greater the pressure, the more focused people become on short-term measures to shore up their indicators instead of activities that may benefit the corporation's greater interests. If sales people, for instance, have to meet a high quota of sales per month, then they are likely to think of nothing but personal survival. In such conditions they are unlikely to pass on clients who may benefit other parts of the organization. If (say) total number of sales is the only criterion, then they are likely to focus only on products they know they can sell and ignore the rest.

In my home state, a controversy arose when the media discovered that Police patrolling the roads had quotas for booking drivers for speeding. The fear was that finding speeding drivers had become their sole concern, rather than overall road safety.

In scientific research, the use of KPIs has greatly warped the way some scientists do their work. The nature of most specialist areas of research has become so technical, so esoteric, that even colleagues working in the same general area may not understand each others work. This makes it difficult for committees to assess how good a researcher is when making appointments or allocating funding. They avoided the problem simply by counting the number of articles each scientist had published.

The pressure to "publish or perish" warped the way scientists worked. Many shifted their focus from studying important issues to look instead at ones that yielded quick and dirty results. It warped people's behaviour, and their integrity. At afternoon tea one afternoon, a colleague (notorious for turning each study into a long series of "short notes") asked me about a problem in his latest experiment. I was shocked the next day when he came and apologized for not including me as a co-author on the manuscript he had just submitted for publication, but assured me that he had acknowledge my help. I was shocked that he had seen fit to publish what I considered to be an absolutely trivial result. Another side-effect of the drive for productivity was its effect on research students. In many labs, researchers became worker bees busily generating discoveries for which their supervisors claimed credit.

Recognizing the problem, the research community has tried to turn to indicators of quality rather quantity. But here again, the use of numbers warps behaviour. One indicator of scientific quality is the number of published articles that cite a particular piece of work. But in some cases, this led to the formation of citation circles in which groups of authors would automatically cite each other's work. In the latest attempt to enforce quality, governments are now ranking journals as A, B or C. This immediately created a climate where anything not published in a rank A journal is considered poor research. While rankings are killing off some poor quality journals and publications, they also risk stifling leading-edge research in new areas that fall between traditional specializations.

For centuries, education systems in many countries have used examinations to assess the quality of their students. Almost invariably, the results are expressed as numbers between 0 and 100, or else as aggregated into bands using the letters A to F. In many parts of the world, achieving high numbers in examinations is crucial to a student's future. A high mark can be a ticket to a good career and an easy life; failure can mean a future of disappointment, compromise and struggle. The side effect of this system is extreme pressure on students. In Japan, many kids resort to suicide, either from the fear of failure, or from the shame of failure. But how meaningful are the exam results in reality? In most examinations, the pass mark is 50. But how good really is a professional who possesses only half the knowledge needed by their discipline? Even worse, almost every student "swots" over their books before an examination and promptly forgets most of what they learned afterwards.

## Measuring the Worth of Things

The human tendency to latch onto to a convenient measurable variable has unfortunate consequences when it diverts attention away from the real nature of the processes involved. This problem is illustrated by the controversial issue of measuring human intelligence.[3]

Intelligence is a loose term to describe an amalgam of diverse mental skills and processes. In 1905 French psychologists Alfred Binet and Théodore Simon introduced an intelligence test to help identify children suffering mental retardation. Using the average scores achieved by children of different ages their scale defined an intelligence quotient (IQ) to be 100 times a child's "mental age" divided by the actual age. In 1916, American psychologist Lewis Terman modified the scale to measure intelligence of adults as well.[4] This extended the use of the measure to weed out "feeble-mindedness" among adults.

IQ scales have remained controversial ever since. We saw some of the problems associated with them in Chap. 6. Their inadequacies are most telling in cross-cultural and inter-racial studies. Intelligence tests are culturally bound. They involve interaction and communication between subject and experimenter, assuming fluency and familiarity with certain skills (e.g. language) and with the sorts of material in the test (e.g. numbers, diagrams). Such problems confound and cloud emotionally-charged issues, such as the question of genetic dependence of I. Q., that arise from cross-cultural studies.

Almost certainly the most widespread cultural straightjacket associated with the use of number is to measure the value of things in terms of money. Just as corporations need metrics when they grow too large for people to know everyone, so money makes it possible to measure and compare the worth of things. Like all the other examples of metrics, measuring the value of things in terms of money has influenced society far beyond its basic, practical convenience. Its dominance has often led people to try to measure the value of everything in terms of the monetary value.[5]

For now, the following example can serve to illustrate how misleading the use of a single number can be when we try to measure social progress in terms of money. In a brilliant speech during his ill-fated 1968 presidential campaign, Robert Kennedy spelled out the dangers of blindly measuring America's social progress through a single number—the Gross National Product:

> Too much and for too long, we seemed to have surrendered personal excellence and community values in the mere accumulation of material things. Our Gross National Product, now, is over $800 billion dollars a year, but that Gross National Product—if we judge the United States of America by that—that Gross National Product counts air pollution and cigarette advertising, and ambulances to clear our highways of carnage. It counts special locks for our doors and the jails for the people who break them. It counts the destruction of the redwood and the loss of our natural wonder in chaotic sprawl. It counts napalm and counts nuclear warheads and armored cars for the police to fight the riots in our cities. It counts Whitman's rifle and Speck's knife, and the television programs which glorify violence in order to sell toys to our children. Yet the gross national product does not allow for the health of our children, the quality of their education or the joy of their play. It does not include the beauty of our poetry or the strength of our marriages, the intelligence of our public debate or the integrity of our public officials. It measures neither our wit nor our courage, neither our wisdom nor our learning, neither our compassion nor our devotion to our country, it measures everything in short, except that which makes life worthwhile. And it can tell us everything about America except why we are proud that we are Americans.[6]

## End Notes

[1]From *The Art of War* by Sun Tzu, Chap. VI, paragraph 14. Translated and edited by Lionel Giles 1910 [Project Gutenberg E-text #132, 1994].

[2]This is discussed further in Chap. 15.

[3]Fancher (1985).

[4]Terman's scale used the statistical distribution of all test scores. It assigned an IQ value of 100 to the average score and the IQ score of individuals was calculated using the formula:

$IQ = 100 + 15 \times$ *(score–average)/standard deviation.*
[5]This is discussed further in Chap. 18.
[6]Robert Kennedy, March 18, 1968, extract of a speech given at the University of Kansas: http://en.wikipedia.org/wiki/Robert_F._Kennedy.
Kennedy was assassinated during the campaign.

# One Thing Leads to Another

**9**

*For want of a nail, the shoe was lost.*
*For want of a shoe, the horse was lost.*
*For want of a horse, the rider was lost.*
*For want of a rider, the message was lost.*
*For want of a message, the battle was lost.*
*For want of a battle, the kingdom was lost.*
*And all for want of a nail.*[1]

### Abstract

Sometimes one event leads to another, forming a chain of causation that produces unanticipated consequences. During travel, a small initial delay can grow to result in very late arrival. Government action to improve safety can lead to increasingly complex and restrictive rules and to undesirable outcomes, such as the restrictive "nanny state". The creation of interlocking treaties between European nations in the late 1800s meant that a single assassination escalated into World War One and the death of millions.

## A Chain of Events

*7 a.m., Tuesday, Hometown USA.* It is a beautiful spring morning. Birds are singing right outside George's bedroom window. The alarm wakes George just in time for the early news on his bedside radio. He showers and shaves while his wife Norma fixes breakfast. After a relaxed cup of coffee and two slices of toast, George kisses his wife and children goodbye and heads off in plenty of time to catch the 8 a.m. express train to work. The sun is shining. George feels great. Life is a bed of roses.

D. G. Green, *Of Ants and Men*, DOI: 10.1007/978-3-642-55230-4_9, 95
© Springer-Verlag Berlin Heidelberg 2014

*7:05 a.m., Wednesday, Hometown USA.* Next morning, George's alarm fails to go off. Unknown to George, the cat knocked it over the previous day while trying to catch a bird through the open window. Fortunately, George's body is so attuned to his routine that he soon wakes up anyway. But he is five minutes behind schedule. Entering the bathroom, he rushes his routine. In his haste, George nicks himself while shaving. Instead of making up time, he wastes precious minutes stopping the bleeding. Annoyed with himself, George then tries to gulp down his cup of coffee. It is so hot that it scalds his tongue. He ends up splashing half the cup down his front. After shouting at his kids and changing his shirt, George is fifteen minutes late leaving home. He is in such a rush; he does not even kiss his wife goodbye. Fifty paces down the street he realises he has left his office keys behind. He races down the street and arrives at the train station just in time to see the 8:03 express leaving from platform 2. George has no option but to wait and catch the next train, which stops at all stations. By now, George is depressed. He knows that he's going to cop a verbal lashing from his control freak of a boss when he walks into the office half an hour late.

Many bad situations arise from chains of events. As they did for poor George in the above story, each event sets up conditions that make the next event inevitable. Break the chain, stop one event from occurring, and the situation never arises. The problem is that it's not always clear where a sequence of events is heading. In this chapter we look at various ways in which chains of events can form and lead to unexpected, and sometimes disastrous consequences.

As we did in Chap. 7, one way to understand accidents is to look at them as networks of possibilities. Each accident is the end result of a sequence of events that lead up to it. Any particular sequence may be unlikely, but there are so many possible chains of events, that some proportion of them will actually occur. Certain steps in the sequence, such as a driver failing to check his brakes, or heavy rain falling, create a rich variety of new pathways that lead to accidents. If it rains, then many cars will fail to keep a safe stopping distance from the car in front. If it is dark, then people will have trouble seeing. If a car has not been serviced, then its brakes are more likely to be faulty. Each of these factors opens new pathways that could lead to an accident.

Even in bad conditions, the chances are still small that any particular car, on any particular stretch of road at any particular time will be involved in a fatal sequence of events. But given thousands of cars, on hundreds of roads over a long enough period of time, it becomes almost certain that something will go wrong somewhere.

## The Domino Principle

Everyone knows about dominoes. When you stand them up in line, you can push one over and it knocks over the next domino, which then knocks over the next, and so on. Students regularly set up spectacular shows in which thousands of falling dominoes, sometimes covering the entire floor of a large hall, make fantastic patterns as they tumble. The thing to remember about dominoes is that because they are all lined up, it takes only a single domino falling to start a chain reaction in which thousands of dominoes go down one after another.

Many processes are like those falling dominoes. An epidemic, for instance, spreads from person to person. It takes only a single infected person to start an epidemic. Each person who becomes infected passes it on to another person, often several other people. Likewise, in a forest fire, the flames spread from tree to tree. A nuclear explosion is yet another kind of chain reaction in which splitting one atom releases particles that split more atoms.

At a news conference on April 7, 1954, U.S. President Eisenhower likened the way that communism had spread throughout Europe to falling dominoes:

> You have a row of dominoes set up, you knock over the first one, and what will happen to the last one is the certainty that it will go over very quickly. So you could have a beginning of a disintegration that would have the most profound influences.[2]

The fear was that if a communist regime gained control anywhere, then the entire region was at risk of falling under communist domination. In particular, U.S. administrators feared that communism would spread from China into surrounding countries. There was some justification for their fear. In the early 1950s, they had fought a war in Korea to prevent the communist North, which had Chinese backing, from taking control of the entire Korean peninsula. This so-called *Domino Principle* coloured much of U.S. foreign policy during the 1950s and 1960s. Possibly its most significant result was America's disastrous involvement in the Vietnam War.

Many kinds of human interactions are like dominoes. One of the most familiar is gossip. Friends and colleagues stop to chat, spreading the latest social news and gossip. Mention that Mrs. X is going to have a baby and suddenly the entire community knows. As a way of spreading social news, it can be very useful. But it also provides a means by which spiteful individuals can hurt others. Many a life or career has been ruined after gossip has converted a single piece of misinformation into a malicious rumour.

## A Horror Plane Trip

The Domino Principle influences human affairs when chains of events cascade in unpredictable ways. As the following story shows, the results can be as annoying as they are unpredictable.

In February 1994 I was heading for a conference in Brazil. Starting from Auckland, New Zealand, my plan was to fly to Los Angeles, transfer to a flight to Sao Paulo, then catch a bus the last 100 km to the city of Campinas, where the conference was to be held.

As I waited in the terminal at Auckland, everything seemed to be in order when the plane I had to catch touched down. Then the problems began. Just as I began gathering my belongings in anticipation of boarding, an announcement informed us that the plane had struck a bird on landing. There would be a "short delay" while they checked the engine for damage.

It was more than two hours before they let us board the plane. But the plane was not yet ready to take off. It is the only time I have ever known an airline to serve a meal on board, and show a movie, while the plane was still on the ground! Later I learned that the only reason the airline did not postpone the flight until the following day was that there were not enough beds in the city to cope with 350 waiting passengers.

When the plane finally took off, after a five hour delay, we thought that our troubles were over. They were not. This was only the beginning. Because of the long delay at the airport, the flight crew had been on duty too long and had to be relieved. This meant that the plane could not head straight for Los Angeles. So we diverted to Honolulu, where we waited for two hours in a hot and steamy transit lounge while the plane was refuelled and a new flight crew found.

When we finally reached Los Angeles, the flight was over nine hours late. My flight plan provided six and a half hours in transit before my next flight. But because we were so late, my next flight had left long ago. I was not alone. About 300 passengers queued up in the transit lounge needing to arrange their connecting flights. My case proved so difficult that 3 h later, I was the last passenger still there.

Because my Brazil hosts had booked flights for me, the schedule was inflexible. So having missed a connection, I was in trouble. At one point the airline representatives suggested putting me on a roundabout route via six US cities to try to catch up with my flight, which was operated by a different airline. Eventually sanity prevailed and they booked me on one of their own flights. After a few sleepless hours in a nearby hotel I boarded a flight to Sao Paulo via Florida. This last flight turned out to be one of the best I'd ever had. With three quarters of the seats empty I was able to stretch out and catch up on sleep. Next morning I woke shortly before landing in Sao Paulo.

On clearing customs at Sao Paulo's beautiful terminal I was just starting to wonder what my next step would be, when I noticed a man in uniform carrying a placard with my name. He handed me a message to the effect that the driver would chauffeur me to the conference venue. Communication was difficult. He spoke none of the languages I could use, and I spoke no Portuguese. So after some pointing at a phrase book we set off. It was only once we left Sao Paulo behind and passed signs on the highway that I realised the truth. Incredibly, my hosts had arranged a taxi to drive me the whole 100 km to Campinas.

When we finally arrived, I was just in time for lunch. By now the delays had blown out to nearly 48 h. I arrived just in time to attend the last session of a three day conference.

When circumstances blow out like they did on that trip, the consequences are simply unfortunate. But as we will see later, they sometimes create the setting for tragedy.

## From Due Care to the Nanny State

"They ought to do something about it." This is a common expression of complaint. People become annoyed about what seems to be a completely obvious problem and want to see action taken to correct it. Who the anonymous "they" are is usually left unstated, but the implication is that it refers to the government, the local council, or some other official body with the means and authority to take the appropriate action. Over time, dealing with lots of "ought tos" becomes a matter of tweaking a complex system to deal with one fault after another.

Some of the most visible examples of "tweaking" complex systems are seen in the quest for safety. Building regulations, for instance, have evolved over the years to ensure that constructions are as safe as they can be for the people who use them. Vehicle registrations in many countries require safety inspections to ensure that they will not be dangerous on the roads. Health and safety regulations lay down requirements about food processing by distributors and health standards for food preparation in restaurants.

You can get an inkling of the increase in regulation and standardization by looking at any soft drink bottle. Many years ago, a soft drink bottle would show only the product name and the name of the manufacturer. Today, the label on a typical bottle would contain the following required information as well:
- A bar code;
- A table showing a detailed analysis of ingredients;
- A recycling stamp;
- A product expiry ("use-by") date stamp.

Public regulation gets caught in an endless cycle of trying to close barn doors by updating rules and standards. Unfortunately, when you try to close barn doors to solve one kind of problem, you often creates new problems of a different kind.

In the early days of the World Wide Web, universities and other early adopters had difficulty coming to terms with the new medium. On the one hand, administrators wanted academics to promote their organizations by providing information about research and teaching online. On the other hand they wanted to control the content to make sure it promoted their "corporate brand." So they laid down rule upon rule about form and content that all web pages had to satisfy and became ruthless about removing anything that did not conform. In some cases, frustrated academics responded by closing substantial websites they had created. After several years of this regime, administrators could not understand why almost no one was providing content for the institution's web site. In its preoccupation with control, the institution had lost the plot. Only gradually did they learn that to manage an activity effectively, you need both sticks and carrots. If you insist that people follow a particular style, then you need to provide them with tools and support to make the job easy.

Regulations sometimes create more problems that they solve. In my home town, for instance, some local councils used to impose fees on homes that installed water tanks, arguing that it reduced storm water flow. However, this practice only served to accelerate rapidly worsening shortages in the urban water supply.

More generally, regulation imposes costs. First, there is no sense in regulation without enforcement. Then again creating regulations and enforcing them impose costs on the tax-paying public. In some cases, local councils have seen regulation as a profitable source of revenue. For suppliers of services, however, the side effect of compliance with regulation has been to create extra costs.

One of the dangers in confusing extreme events with means is that they can provoke over-reaction. Knee-jerk responses to extremes have led to a steady erosion of living standards or personal freedom.

The tightening of airline security, for instance, has been a history of imposing new airport checks and restrictions after every incident. Plane hijacks by terrorists are rare events, but the fear of them has driven governments and airlines to treat every passenger as a suspect. During the 1960s and 1970s, airline hijacks made frequent news headlines. To counteract this, airports introduced security screening. To foil the criminal extreme, treating passengers as suspected criminals became the norm. In 2009, a single incident in which a passenger smuggled explosive liquids on board a flight in the U.S. was enough excuse to provoke a massive increase in security screening at airports.[3] Carrying liquids, any liquids, on board a plane is now forbidden in the U.S. Full body scans at airports are now becoming common in the U.S. and elsewhere. The result is that air travellers are now subjected to continual annoyance, frustration and delays. This is a minor victory for the terrorists.

Trying to prevent extremes by introducing new rules and regulations is a bit like patching holes in a boat that is constantly springing new leaks. In many areas of life, the result has become a mountain of rules, regulations, standards, certifications and licences, not to mention reporting, testing and auditing requirements. It has led to local governments introducing a plethora of local rules and regulations. This raised concerns about the so-called "Nanny State." Like a nanny caring for an infant, the nanny state tries to protect its citizens (and itself) by binding them in a tangle of ever more restrictive rules. In my home town, street parties used to be a popular community activity: neighbours all contributed food and drinks and gathered on the street to get to know one another. At the time of writing this, however, to hold a street party you first need to obtain permits from local council, police, and emergency services, as well as obtaining public liability insurance. You also need to satisfy regulations about food preparation, responsible supply of alcohol, and safety, to mention just a few. No wonder that most people, faced with such a mountain of restrictions, simply give up and ignore their neighbours.

## Armageddon

There are times and places in the course of human affairs when the fate of nations hinges on the turn of a single event and when every action is pregnant with its consequences. One such moment occurred during the summer of the year 1914. Peace had reigned in Europe for almost a century.[4] There had been no widespread conflict on the continent since the defeat of Napoleon in 1815. The Franco-Prussian War of 1870 had been a short, sharp affair. Other wars between European nations had been fought on far-flung battlefields, such as the Crimea.

For ordinary citizens in a dozen countries, it was an unrivalled era of peace and prosperity. Nevertheless, the seeds of Armageddon had long since been sown. Over a period of decades, growing nationalism and economic rivalry had led to a vast build up of military power all over the continent. It needed only a spark to set off the powder keg.

There was a festive mood in the town of Sarajevo on the morning of June 28, 1914. The heir to the thrones of Austria and Hungary, Archduke Franz Ferdinand, was attending army manoeuvres just outside the city. At 10 a.m., the Archduke and his wife Sophie left the army camp and drove to the city hall along the Appel

Quay in a Graf and Stift motorcar. As the motorcade rolled past flags, flowers, and cheering crowds, their driver saw an object tossed from the crowd and accelerated to avoid it. The bomb bounced off Franz Ferdinand's arm onto the pavement. It exploded just feet away. Naturally the royal couple were badly shaken by this incident, but they regained their composure during a reception at City Hall.

Despite the evident danger, they had no hesitation in setting off on the return journey. They had not gone far when their driver took a wrong turn and the car was forced to halt at an intersection, just feet away from a Serbian nationalist named Gavrilo Princip. Seeing an opportunity, the assassin stepped forward from the crowd and shot them both. Sophie died on the way to hospital. Archduke Franz Ferdinand died soon after being admitted.

Terrible as it was, this assassination sparked off a chain reaction of events that were infinitely worse. For several weeks after the assassination, it seemed as if nothing would happen. However, this was only because it took time for the protagonists to mobilize their vast armies. At no stage in what followed did anyone anticipate the horrors to come. At no stage did anyone intend to start a World War. And yet that is exactly what happened.

In hindsight, it was almost inevitable. Over the years, countries all over Europe had set up a complex network of treaties and agreements with one another. Using the assassination as an excuse, the Austrian government declared war on Serbia. Invoking its defence agreements with Austria-Hungary, the German government followed suit and went to war also. However, following the Franco-Prussian War, German strategy for general war had been based on what was known as the Schiefflen Plan. This involved invading France and outflanking its army by making a great wheeling movement through Belgium. Implementing this plan set off a cascade of agreements that brought more and more countries into the conflict, including Britain, Russia and their allies. The following escalating sequence of events then took place:

| | |
|---|---|
| 28 July | Austria declares war on Serbia |
| 1 August | Germany declares war on Russia |
| 3 August | Germany declares war on France |
| 4 August | Germany invades Belgium |
| | Britain declares war on Germany |
| 5 August | Montenegro declares war on Germany |
| 6 August | Serbia declares war on Germany |
| | Austria declares war on Russia |
| 12 August | Britain and France declare war on Austria |
| 23 August | Japan declares war on Germany |

By the outbreak of fighting, millions of soldiers were on the move. The mood was jubilant as each country sent its troops off to what they thought would be a short, glorious campaign. Recalling the swiftness of the Franco-Prussian War, over forty years earlier, everyone expected it to be over by Christmas. When the Great

War did end, more than four years later, its impact had been truly terrible. Countries on every continent had been involved in the conflict. Over 65 million people had been mobilized. Of those, 8.5 million had been killed and a further 29 million were wounded, missing or prisoners of war. The governments of Germany and Russia were overthrown.

The repercussions of the Great War went on long after the fighting ended. In the Middle East, for instance, Arab tribes were drawn into the fighting, setting the scene for nationalist movements in later years. Ultimately, the Great War, later known as World War 1, sowed the seeds of World War 2 and the Cold War. It also led to revolution in China and to independence struggles in dozens of countries around the globe. It reshaped the world in ways undreamed of by those involved at its beginning. For the most part, they would have been horrified had they known even half of the consequences of their actions.

And in the end what was it all for? A lot of reasons, and some justifications arose only after the fact. The tragedy is that at any time the escalating sequence of declarations above could have been stopped if only the leaders could have stepped back and said to each other: "This is madness, let's stop now." But as the British philosopher Bertrand Russell aptly put it:

> And all this madness, all this rage, all this flaming death of our civilization and our hopes, has been brought about because a set of official gentlemen, living luxurious lives, mostly stupid, and all without imagination or heart, have chosen that it should occur rather than that any one of them should suffer some infinitesimal rebuff to his country's pride.[5]

## End Notes

[1]Nursery rhyme, attributed in part to Benjamin Franklin.

[2]Public Papers of the Presidents, Dwight D. Eisenhower, 1954, pp. 381–390.

[3]In 2009, Abdul Farouk Abdulmutallab smuggled 80 g of liquid explosive on board a flight bound for Detroit. http://www.independent.co.uk/news/world/americas/wealthy-quiet-unassuming-the-christmas-day-bomb-suspect-1851090.html.

[4]There are many accounts of the events leading up to the start of World War I. The summary presented here is based on Taylor (1963).

[5]Letter to the editor of "Nation" from The Collected Papers of Bertrand Russell by Bertrand Russell, (ed. Richard A. Rempel) published by Routledge, 2003. ISBN 0415104637, 9780415104630, pp. 774.

# The Snowball Effect

**10**

*Expansion means complexity, and complexity means decay.*[1]

## Abstract

Events sometimes form cycles in which the outcome of one event feeds back into itself, leading to ever bigger consequences, or "positive feedback". Examples include cycles of revenge between rival groups, growth of compound interest, and the formation of market price "bubbles". In contrast to negative feedback, which dampens change, positive feedback sends systems out of control. It converts small, local differences into large, global patterns. Examples include the way real estate prices soar in some areas, and the way some media performers rise to become "superstars". Positive feedback also played a part in increasing levels of sugar and salt in processed foods and a resulting rise in the incidence of diabetes and heart disease in western countries.

## Snowballs and Bank Accounts

Everyone knows about snowballs. The idea is that you start at the top of a hill covered in snow and let go of a small snowball. It rolls down the hill and gathers more snow on the outside. By the time it reaches the bottom, the small ball has grown into a much larger one. This tendency to grow faster and faster is often called the *Snowball Effect*.

There are many examples of the Snowball Effect, both in nature and in human affairs. Your bank balance is one of them. Put $100 in a bank savings account that earns 10 % interest per annum and after 5 years the balance has grown to $161. The more there is in the account, the faster it grows, so after another 5 years, the balance has grown to $259. This is a well-known pattern of increase is called *exponential growth*.

D. G. Green, *Of Ants and Men*, DOI: 10.1007/978-3-642-55230-4_10,
© Springer-Verlag Berlin Heidelberg 2014

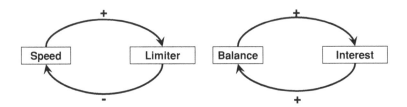

**Fig. 10.1** Examples of feedback. *Left* a control system such as a speed limiter provides negative feedback: if speed increases, the limiter decreases it again. *Right* Compound interest forms a positive feedback: if the bank balance increases, interest increases and this further increases the bank balance

Underlying the Snowball Effect is a well-known process called *feedback*. Feedback means that the results of some process feed back to influence its course in the future. The idea of feedback arose from the need to control machinery. For example, governors on steam engines and thermostats on ovens are control devices that form *negative feedback loops* with the equipment they control. These loops ensure that crucial properties (speed and temperature in the example given) remain constant. It was soon realised that similar mechanisms also existed in nature. The human body, for instance, has control systems to regulate many functions, such as heartbeat and body temperature. In 1948, Norbert Wiener coined the name *cybernetics* to describe the science of control and communication systems in both animals and machines.[2]

A common aim of control theory is to maintain a system in a steady state, or *equilibrium*. A system in equilibrium does not change of its own accord. Control theory identifies two kinds of feedback. In *negative feedback*, the process acts to counter any change, so promoting a stable equilibrium state (or oscillations). The best-known examples of negative feedback are the regulatory systems that we saw above (Fig. 10.1).

In contrast, *positive feedback*, acts to continue and increase any change, so promoting instability and rapid movement of a system away from its initial state. Positive feedback is responsible for the Snowball Effect. As we saw above, compound interest is a common example (Fig. 10.1). Sometimes there are interlocking feedback loops (Fig. 10.2).

Traditionally, positive feedback has been seen as a bad thing. It does, after all, make systems unstable, and that is exactly what control engineers try to avoid at all costs. However, (to use a pun) there is a positive side to positive feedback. As we shall see below, positive feedback—the Snowball Effect—plays an important role in creating order within systems.

Positive feedback, and the exponential growth that accompanies it, is a characteristic of complexity. Small, local differences do not even out over time, they become more pronounced and can grow to become large-scale patterns. Whether it is an assassination, a missed volley in a tennis match, customers racing to buy a new product or patches of clover spreading to take over your lawn, local events

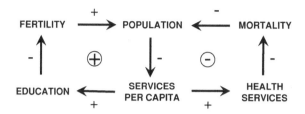

**Fig. 10.2** An example of interlocking feedback loops. The *arrows* indicate the effect of one variable on another. Mortality, for instance, has a negative effect on population size. The circled signs indicate the overall effect of each feedback loop. This figure elaborates a segment of the limits to growth model (Fig. 3.1)

can have global consequences. There are many everyday examples of the exponential growth patterns that result. In the following sections we look at some of them.

## An Eye for an Eye

At the start of the 18th century, a Japanese courtier named Asano Naganori became embroiled in an argument with a court official named Kira Yoshinaka. Blinded by anger, Naganori assaulted Yoshinaka. In punishment he was obliged to commit suicide, leaving all his samurai retainers as masterless ronin. Banding together these forty-seven ronin plotted and schemed patiently for 2 years. Eventually they succeeded in killing Lord Yoshinaka in revenge for the death of their lost master and then committed suicide themselves.[3]

Throughout history, revenge has been a powerful motivation for human hatred and violence. Driven by unbearable loss, a sense of incompleteness, the need to act, any crime against an individual close to you demands repayment in kind. Revenge is a horrible example of positive feedback in a social setting.

In the play Romeo and Juliet, Shakespeare portrays a typical case of revenge spiralling out of control. A careless remark by the leader of the house of Capulet prompts his opposite number in the house of Montague to seek revenge for the slight against his honour. This action leads to an escalating cycle of revenge in which each offence raises the desire for retribution on an ever-increasing scale. By the play's opening, hatred between the two noble families has grown to the extent of igniting a third civil brawl. This leads the prince to threaten the two patriarchs with death for any future violence, and sets the scene for the star-crossed lovers from the two warring families.

So it is that one crime leads to another, and another … and another. Burning desire for revenge often leads to an endless cycle of hate and violence. Cycles of revenge are well known throughout history. They occur in many settings: from tribal warfare in New Guinea to gang warfare in American cities.

## From Ants Hills to Town Planning

The Snowball Effect plays an important part in creating order and pattern in complex systems. Let's look at what happens in an ant hill. In the book *Patterns in the Sand*, I described how the organization of an ant colony emerges without any conscious planning on the part of the ants. The order emerges out of the activity of hundreds of ants interacting with their environment. As we saw in Chap. 2, all the ants do is to obey certain simple rules. For instance, if an ant comes across a piece of rubbish, it will pick it up. If it is carrying a piece of rubbish and comes across a pile of rubbish, then it will drop the piece that it is carrying.

Now, there is a potential problem with this strategy. The above behaviour produces not one pile of rubbish, but many. So how do the ants avoid becoming stuck in an infinite loop, endlessly transferring rubbish from one pile to another? Fortunately, positive feedback comes to their rescue. What happens is that, purely by accident, some piles will be slightly larger than others. This difference is all-important. It means that those piles will gain scraps faster than they lose them.[4] Each scrap they gain increases the advantage of the big piles over smaller ones. As time goes by, the smaller piles shrink, becoming smaller and smaller. The rate at which they shrink increases rapidly. They soon collapse and disappear altogether. The big piles continue to gobble up scraps from smaller ones. Eventually, there is just one pile left.

This process may seem very inefficient. In the early stages it is. But with hundreds, or even thousands of ants working on the problem, it all happens very quickly. And it all happens with no overall plan, no overall design.

In the ant colony, positive feedback leads to a simpler, more workable organization. However, in human cities, knee-jerk decisions can lead to feedback effects that make problems worse instead of better. As cars became more common during the mid-twentieth century, narrow urban streets became clogged with motor traffic. In many cities, town planners responded by building wider roads and freeways. The availability of better roads led families to move out of the crowded inner city to the suburbs and rely on their cars to commute to work each day. Town planners responded to the increase in traffic in the outer suburbs by building more roads and freeways. Thus families and planners became caught in a feedback loop: having roads to use encouraged people to use more cars and more cars led planners to build more roads. So we have created sprawling cities with people driving long distances and creating huge traffic jams as they converge on town centres.

## Location, Location, Location

We can see positive feedback at work in the ways towns and cities sometimes organize themselves. Some years ago, the town where I lived had three video stores; each was located in a different part of town. Naturally we borrowed movies from the store nearest to us. But sometimes, when a new movie was in great demand, or when we were looking for something out of the ordinary, we would do a tour of the stores until we found what we were looking for. Then one day, the local video store closed and shifted downtown. One of the other stores did the same thing. Suddenly all the stores were located not just in the same area, but in the same street. Two of them were actually sited opposite one another.

The reason for this gathering of similar stores is not hard to see. Almost everyone went shopping downtown at some time of the week, whereas only locals would go to the edges of town. So the store that first located itself downtown was doing better business than its competitors. Not surprisingly, these competitors saw that the first store was doing better business, so independently they each decided to relocate. And not merely relocate to the down town area. By relocating to the same street as the first store, they assured that they would be in direct competition. That way they had the opportunity to steal customers away from the store that was already there. For customers it meant that the street concerned became the video district of town, the place you went when you wanted to rent a movie.

In some cities, positive feedback has taken hold and led to specialization of entire quarters. In Tokyo, for instance, the suburb Akihabara became known as Electric Town because of the high concentration of specialist electronics stores.

In our town, however, the arrangement did not last. The town where we lived was not large enough for three video stores to thrive and after a few years only one store remained. Unfortunately for us, the winner was the store that had the poorest range of titles. It won by restricting its stock to the best selling titles and by using volume to keep its prices low.

On a small scale this story highlights a general principle. Businesses that are successful attract competitors. And those competitors often locate themselves close to the original business for two good reasons. First, the original business often set up in an area that best assured success. Secondly, by locating close to the first business, they give themselves a much greater chance of securing a share of the market because customers already associate that area with their product.

Incidentally, similar principles extend to the Internet as well. The Internet overcomes geographic constraints on selling. It makes it possible for (say) manufacturers to sell their products direct to customers. However, if you have wanted to offer your product online, then there are considerable advantages in making it available through a site that already offers similar products. This has led to a proliferation of integrator sites that, like traditional retail outlets, offer a wide range of products from many specialist suppliers.

## The Big Get Bigger

The Snowball Effect not only plays a role in determining where businesses are located but, also, whether they succeed or fail in the face of competition. This is because of the way competition can affect product sales. For instance, during the 1980s, many companies jumped on the bandwagon of a new technology and set themselves up to sell personal computers. Most of these companies soon went out of business or were gobbled up by competitors.

Large companies have many advantages over smaller ones. Not only do they tend to have more capital, they also benefit from economies of scale. So when a large company comes into competition with a small one, odds are that the large one will win. So big companies tend to get bigger and swallow up smaller companies.

Advances in communication have changed the playing field in which business operates. Given their imperative to grow, companies have to exploit communications to find new markets and to become more efficient.

What is happening is typical of many complex systems. Change the nature of its internal workings and the system reorganizes itself. In this process, most companies will fall by the wayside or be gobbled up by the select few that grow and grow to become international giants.

Money begets money. The Snowball Effect can lead to a runaway trend. This effect plays a prominent role in globalization. Large organisations tend to get larger. For example, if a company succeeds in taking over a competitor, then it acquires resources and capital that enable it to grow still further. Likewise, if it amalgamates with (say) one of its suppliers, then it gains the potential to grow by diversifying into new areas. In contrast, if a small company experiences setbacks, then it becomes harder to bounce back and start growing. What follows can be a vicious circle, a positive feedback loop, in which one problem leads to another, until the company folds, or gets swallowed by competitors.

When the Internet first became seriously commercial during the 1990s, literally thousands of Internet businesses sprang up, virtually overnight. What followed was a process of intense competition. Companies that were well-organized, well-placed, or simply lucky, flourished and grew. They rapidly gobbled up other start-up companies. Within a few years the marketplace had resolved itself into a relatively small number of large players.

The same process acts on entire communities. Small towns have trouble attracting businesses, which prefer the advantages offered by large cities. The lack of opportunity leads people to leave their small town or village and head for the bright lights. This movement accelerates the process. The local bank closes its doors. Businesses fail. Small towns all but die. In Brazil for instance, I have seen whole shanty towns that sprang up around the outskirts of cities, as poor families flocked in from the countryside seeking work. The same story, with lots of variations, repeats itself in dozens of countries.

## The South Seas Bubble

In 1711, the British government was facing a crisis. It had accumulated a huge debt of about ten million pounds sterling from its part in the War of Spanish Succession. Aiming to restore public confidence, a group of merchants, led by the

Earl of Oxford, took over the debt. The government agreed to pay them 6 % annual interest. To meet this, it raised duties on wine, tobacco, vinegar, silk and various other imported goods. It also granted the group a monopoly on trade to the South Seas and incorporated them under that name by an act of parliament.[5]

One of the prizes the South-Sea Company sought to exploit was the seemingly inexhaustible wealth emanating from gold and silver mines in Mexico and Peru. At the conclusion of the War of Spanish Succession in 1713, the Treaty of Utrecht granted Britain trade access to Spain's South American colonies. The South-Sea Company obtained a contract to supply slaves from Africa in return for limited trade with Mexico, Peru and Chile. This trade, however, was subject to an exorbitant 25 % tax on the profits. The Company's first annual ship set sail in 1717, but a rupture with Spain halted trade the following year.

In his speech to open parliament in 1717, King George pointed to the need to reduce national debt. On 20th May 1717, the Bank of England and the South-Sea Company each put forward proposals, which were accepted by parliament. The South-Sea Company reduced its interest to 5 % per annum in return for being allowed to increase their capital to 12 million pounds by subscription.

At about this time John Law's Mississippi scheme (next section) was inflaming greed among the population of Paris. Seeing this ...

> ... inspired them (the directors) with an idea that they could carry on the same game in England ... Wise in their own conceit, they imagined they could avoid his faults, carry on their schemes for ever, and stretch the cord of credit to its extremist tension, without causing it to snap asunder.[6]

Upon approval of the company's scheme by Parliament on 2nd February 1720, London went crazy. The value of the company's stock rose rapidly, even while the bill was being debated. From a price of £128 per share in January it rose to £330 in March.

The success of the South-Sea enterprise inspired numerous other schemes "... *each more extravagant and deceptive than the other.*"[7] These "bubbles", as the schemes soon became known, often lasted no longer than the time it took for the perpetrators to sell a body of stock and cream off their profits. On 17th July 1720, the Board of trade dismissed petitions for some 86 such bubbles, making them illegal. However speculative schemes continued to flourish nevertheless.

Meanwhile the South-Sea Company continued to increase in value. By 28th May its stock had risen to £550, and then leapt to £890 by the start of June. At this point, many investors began to feel that the stock could rise no higher and began to sell. Buying by the directors steadied the price, which stood at £750 on 22 June. Various strategies by directors restored confidence and the price gradually rose to £1,000 by the beginning of August.

The bubble was now so inflated that it had to burst sooner or later. Information that the chairman Sir John Blunt had sold out, along with some other directors, led to general uneasiness and the price fell to £700 by 2 September. With public alarm mounting the directors summoned a meeting of investors on 8 September. However, this had little effect and the price continued to freefall. By 12th September, the price had fallen to £400 and by the end of September was just £135. The Bank of England tried to intervene and save the situation, but pulled back when it became clear that collapse of the bubble could ruin them too.

Innumerable investors were ruined in the collapse. Anger directed against the company directors was intense. A parliamentary enquiry found evidence of fraud, bribery and profiteering. Several people were brought to trial and several convicted. Most notable of all was John Aislabie, the Chancellor of the Exchequer, who was found guilty of corruption. Aislabie had received a bribe of £20,000 in company stock in exchange for his support of the scheme. He was convicted and imprisoned in the Tower of London.

The South Seas affair led to lack of public confidence, especially in foreign investments, that took many years to restore.

## Mississippi Madness

Another infamous bubble was the madness over Mississippi that overtook France in 1719 and 1720. It was fuelled by the activities of a single individual, a Scotsman named John Law.[8] Born in Edinburgh in 1671, Law learned the principles of finance by working in his father's counting house. When his father died in 1688, he quit Edinburgh and moved to London where he lived for many years off his profits from gambling.

He was forced to leave London after being sentenced to death in London for killing the husband of his mistress in a duel. He escaped and made his way to the continent. After arriving in Paris in 1716, he presented himself at the royal court. On the strength of his writings on finance, he was granted permission to establish a bank on 5th May 1716.

By ensuring that his bank notes were redeemable at the current face value, his reputation, and fortune, grew rapidly. Confidence in his bank boosted public commerce. Within a year he had to open branches of his bank in five other cities.

Law's astonishing success so impressed the regent, Philippe d'Orléans, that he approved a new project in which Law would have exclusive rights to trade along the Mississippi River. The new company was incorporated in August 1717 and 200,000 shares were issued at 500 livres each. Based on Law's good reputation and the confidence inspired by the performance of his bank, people flocked to the scheme.

It was at this point that John Law and the regent together made their fatal mistake. Both of them appeared to forget Law's own maxim that banks should not issue paper for funds they did not possess. Intoxicated with the success of the Mississippi scheme, the regent fabricated bank notes to the amount of one billion livres. When the chancellor D'Aguesseau tried to oppose the flood of paper money, the regent dismissed him from office. When parliament tried to ban paper money as acceptable payment, the regent overruled their decree. At one point Law had to seek the protection of the regent when members of Parliament proposed that Law should be set on trial.

Meanwhile the value of the shares continued to rise. Early in 1719, the regent granted the Mississippi Company exclusive right to trade in the East Indies, China, and the South Seas. In response, the Mississippi Company changed its title to Company of the Indies and created more shares, promising a yearly dividend of 200 livres for each 500-livre share.

An enthusiastic public rushed to join the scheme: more than 300,000 applied for the 50,000 shares on offer. With such huge demand people were willing to pay ever higher prices to obtain the shares. A frenzied market for the shares led market price to increase at astonishing rates. In one case a shareholder was ill and sent his servant to sell 250 shares at the most recently quoted price of 8000 livres. Finding that the share price had risen to 10,000 livres, the servant paid back his master the quoted value and absconded that same evening, leaving the country with a fortune of 500,000 livres in his pocket.

Such a hugely inflated bubble could not grow forever. In May 1720 the council of state became alarmed when it was told that the market value of all notes then in circulation was more than twice the total money in the country. On the 21st of May a decree was issued stating that the value of shares in the company and notes from the bank would gradually be devalued to one half their nominal value. On the 27th May the bank stopped payments and Law was dismissed from the ministry. A further series of edicts eventually made the shares and notes worthless.

The effect of these changes on the economy and the people were dramatic. People who were rich one moment suddenly found themselves paupers the next. The anger of those who lost their savings was intense. The regent tried to put the blame on Law. From being the idol of the populace, Law now became an object of hate. When he foolishly attended the opera his carriage was set upon by a mob. Only fast action by his coachman and servants, who closed his gates as the coach entered, saved his life. Shortly afterwards he was forced to seek safety in the regent's palace and some months later was allowed to leave the country. John Law died in Venice in 1729.

---

## The Land of Milk and Honey

Market forces often combine with feedback to create trends in commercial products. If you want your product to sell, then you give people what they want. Take the case of sugar. Humans have always had a sweet tooth. One way to sell a food product is to make it sweet. So you give them sugar, even if they do not want it; even if it is not what they need; even if it is not good for them.

Feedback pushes the trend even further. Over a period of time, people (especially children) develop a taste for sweet products. If people like sweetening in one food, they are likely to prefer sweetening in other foods. So if one company increases its sales by adding sugar, then its competitors will be forced to follow suit or risk losing their market share. This imposes a form of selection that favours sales of sweetened products over unsweetened alternatives. It has also led to increasing amounts of sugar in some products.

The above cycle has led to sugar being added to more and more food products. Some children's breakfast cereals are packed with sugar. Looking along the shelves of our local supermarket shelves I found that popular breakfast cereals currently contain about 10 % sugar. The proliferation of sweetened products has led to very high intake of sugar:

> Annual per capita sugar consumption is now highest in its places of production, such as Brazil, Fiji, and Australia, where it exceeds 50 kilograms.[9]

Sugar intake is only slightly lower (30–40 kg) in western Europe and North America.

The case of salt is similar. Salt enhances the flavour of food. So if you want a food product to be tastier, just add salt. Inevitably, salt addiction created a feedback loop similar to the one for sugar. Most processed food products now contain salt.

Inevitably, intake of sugar and salt in excessive amounts has led to side-effects. Combined with increases in the fat content in foods, and decrease in exercise, they contribute to a wide variety of health problems, especially in the developed world. According to the World Health Organization (WHO):

> In 2008, more than 1.4 billion adults were overweight and more than half a billion were obese. At least 2.8 million people each year die as a result of being overweight or obese. The prevalence of obesity has nearly doubled between 1980 and 2008. Once associated with high-income countries, obesity is now also prevalent in low- and middle-income countries.[10]

In the United States, the number of obesity cases doubled in 25 years, from 15 % in 1976 to 31 % in 2001.[11] The above increases are certainly not the result of sugar and salt alone, but they have contributed significantly.

Another side effect is the increase in the incidence of diabetes.

> Globally, 44 % of diabetes, 23 % of ischaemic heart disease and 7–41 % of certain cancers are attributable to overweight and obesity.[12]

Until the late Twentieth Century the above trends went unchecked. However, increasing awareness of health issues in western countries has forced manufacturers of processed foods to reduce sugar and salt content of some products. Pressure has even led to feedback in the reverse direction: low sugar, low salt and low fat have become marketing labels for a healthy lifestyle. Improving diet is part of the WHO's Global Strategy on Diet, Physical Activity and Health, which aims to:

> … establish and strengthen initiatives for the surveillance, prevention and management of NCDs (Non-Communicable Diseases).[10]

## Winners and Losers

*Australian Open Tennis, Friday 18 January 2002.* It is the second round of matches in the Ladies Doubles competition of this Grand Slam event. The match under way on Court 3 shows a good example of the Snowball Effect at work in human activity. One of the top seeded pairs is playing against two young Australian girls. For a long time, the match is very even. In terms of raw skill, there is almost nothing between the two pairs. Both sides hit plenty of winners, as well as a number of errors. Not a single service game is lost until the score reaches 5 all. Then a couple of loose points allow the more experienced pair to break through and break serve. Nerves may have been at work. To spectators it looks as if the increasing tension has gotten the better of the Australian pair. They do not quite believe they can beat the seeded pair. After being so close, losing the first set rattles them. They proceed to drop serve again at the beginning of the second set. From then on, the result is never in doubt. Their game crumbles rapidly. Each lost game erodes their confidence still further. By the end of the match, they are hardly able to win a point. The final set goes to the seeded pair 6 games to 0.

This manifestation of the Snowball Effect plays a prominent part in sports of all kinds. In tennis, again, it is especially visible. Not just within a match, but you see it in action all through a tournament. In real terms, there is not a huge difference in ability between the top players and lower ranked ones. The real difference is psychological. A top ranked player expects to win. As often as not, when (say) a player ranked 100 loses a game to someone ranked in the top 10, he or she begins to doubt that they can stay with their opponent. So their game goes off a bit, they lose another game. Then their confidence crumbles and it's another win for the champion.

The same ebb and flow of confidence also occurs on the battlefield. In classical times, it was often the morale of the men fighting that won battles. Fighting frequently degenerated into a brutal shoving match. The two armies would hack at each other in a wholesale slaughter until one side cracked. All it needed was for doubt to overcome determination in a few individuals. Panic would then spread quickly. In a matter of seconds, what had been an equal sided contest could degenerate into a rout.

## Why Do Some Guys Have All the Luck?

Behavioural biologists have found that when it comes to sex, success builds on success. In some species, such as grouse, the males gather at certain spots during the mating season and display for the females, hoping to attract a mate.[13] However, what tends to happen is that the females not only prefer males with large territories, but also males who have also attracted other females. The result is that a small proportion of the males end up mating with most of the females. Similar behaviour occurs in other species.[14]

Other research has shown that a similar effect occurs amongst humans. It is more pronounced in women than in men. In all species, including humans, the explanation is not that the winners are infinitely superior to the losers, but rather that a kind of "Snowball Effect" (positive feedback) is at work. Once one female has chosen to mate with a particular male, then other females see her choice as evidence that the male in question is worth mating with. Observing the choices made by other females is a convenient shortcut. If there are 100 men at a party, how do you decide which of them are worth pursuing? You don't have time to chat up each one in turn. But if there are one hundred other girls at the party as well, then you can see which men they prefer. This social phenomenon is like dominoes. Once one girl shows interest in a male, he immediately becomes more attractive to other females.

Some years ago, a local TV show carried out an experiment that demonstrated the above idea in practice. Actually, testing the theory was not their real intention. They were just trying to find out the best way to pick up girls! In their experiment, a male reporter turned up at the same nightclub several nights running. On the first night, he tried using various standard pick up lines, but with no luck at all. On the second night, he dressed up in expensive clothes, but still failed to improve his luck. By far the most successful ploy, he found, was to turn up accompanied by a beautiful model. He just sat in the bar chatting with her, and ignored all the other girls. When he came back, alone this time, girls who had shunned him previously were suddenly keen to make his acquaintance. Just how successful he was that night was not reported!

When you think about it, the same sort of thing happens with many forms of popularity. Fame breeds on fame. That's why bookstores put best-selling novels on a special rack near the entrance. Busy people cannot always afford the time to decide for themselves which books are the best to buy. They buy books that lots of other people have already bought. We see the same effect with the top ten on the musical hit parade. Once a tune starts becoming popular it is bound to skyrocket. Other music or performers may be as good. They may even be better, but unless the perception that they are good takes hold, they risk ending up as a box office flop.

The spread of opinion often dominates in the publicity circus. Once the perception of quality takes hold amongst key opinion makers, it spreads everywhere. Perception becomes reality. So it is that a handful of movie stars will be considered "hot" at any one time. Likewise, in almost any sport, a small number of stars get by far the most of the publicity. For the same reason, prospective employers always seek references for people they are thinking of hiring. Someone else's opinion of the candidate saves them a lot of time and trouble.

## End Notes

[1]Parkinson's Third Law (From *Laws and Outlaws* 1962).
[2]Wiener (1948).
[3]For a recent account see Warner (2013).
[4]Large clumps grow because the size of a clump depends on its area, but the rate of loss depends on the length of its perimeter. The ratio of perimeter length to area decreases as a clump grows. So the rate of loss relative to size decreases as a clump gets bigger, but increases if it gets smaller.

[5]This account is based on Mackay (1841).

[6]Mackay (1841), p. 51.

[7]Mackay (1841), p. 57.

[8]Mackay (1841) pp. 1–48.

[9]Kiple, K. F. and Ornelas, K. C. (eds). (2000). The Cambridge World History of Food. Cambridge University Press, Cambridge. ISBN: 9780521402163. 1958 pages. http://www.cambridge.org/us/books/kiple/sugar.htm (section II.F.2).

[10]WHO (2012). *Obesity and overweight*. WHO Fact sheet No. 311. http://www.who.int/mediacentre/factsheets/fs311/en/.

[11]Penn and Zalesne (2007).

[12]WHO (2012). http://www.euro.who.int/en/what-we-do/health-topics/noncommunicable-diseases/obesity.

[13]Rintanaki, P. T., Alatalo, R. V., Hoglung, J. and Lundberg, A. (1995). Mate sampling behaviour of black grouse females (*Tetrao tetrix*). *Behavioral Ecology and Sociobiology* 37 (3), 209–215.

[14]Galef, B. G. (2008). Social influences on the mate choices of male and female Japanese quail. *Comparative Cognition and Behaviour Reviews* 3, 1–12.

# A Deadly Cascade

## 11

*There is a science of simple things, and art of complicated ones.*
*Science is feasible when the variables are few and can be*
*enumerated; when their combinations are distinct and clear.*[1]

**Abstract**

Often, the world around us is more complex than we know. Of necessity, the models we build to explain the world simplify things by omitting many details. These omissions explain why experts are so often wrong. They also underlies serendipity, an important source of discovery in science. Because models cannot always predict the future behaviour of complex systems, we need to examine scenarios instead. Although we can refine models to account for new events and conditions, they will never be perfect, so unexpected trends and events continue to occur. When conditions change, things we disregard can become important and set off cascades of unexpected effects. This process is often involved in major accidents.

## Why Experts Fail

Who is better at making predictions, an expert, or a child picking numbers from a hat? Philip Tetlock, a psychologist at the University of California set out to answer this question. Over a twenty year period from 1985 to 2005, he traced the accuracy of some 28,000 predictions made by 284 experts in economics and politics.[2] His results were both startling and unsettling. They showed that the accuracy of predictions made by the experts was only slightly better than chance. More disturbing still, Tetlock found that the more certain experts were of their predictions, and the

D. G. Green, *Of Ants and Men*, DOI: 10.1007/978-3-642-55230-4_11,
© Springer-Verlag Berlin Heidelberg 2014

more well-known and trusted the expert, the less accurate their predictions were likely to be. And yet, people still believe them.[3]

Tetlock also found that some of the worst pundits were experts classified as "hedgehogs". These are experts who base their predictions on sweeping, general ideals or theories, such as liberalism or Marxism. Perhaps because their certainty inspired confidence in like-minded thinkers, hedgehogs tended to be well-known, influential and trusted. More successful in their predictions were experts classified as "foxes," who based their predictions on a more eclectic range of criteria. Foxes also tended to be more cautious about their predictions, which were often narrower and shorter-term.

The problem of expert predictions reflects a problem in modern science: the growing volume of knowledge leads to increasing specialization. Up until the end of the 19th century, understanding ideas at the forefront of science was within the reach of most intelligent people. During the Twentieth Century, however, the rapidly growing volume of knowledge, and its technical nature, made it increasingly difficult for the lay public to understand modern science. Today, even professional scientists are unable to understand the technical details of ideas outside their own field. The increasing specialization of science led Edward Teller, not entirely tongue in cheek, to suggest that:

A specialist is someone who knows more and more about less and less until he knows everything about nothing.[4]

The trend towards increasing specialization has been driven chiefly by the growing volume of scientific knowledge. It is also encouraged by the reductionist nature of traditional science. Science tries to simplify the problem of understanding complex phenomena by reduction—by breaking them down and understanding the components. The idea is that if you understand the parts, then you understand the whole. Understand how the heart, lungs and other organs work and you understand how the body works. The approach has served science well for hundreds of years. But as we saw in Chap. 8, the divide and rule approach fails to cope when things are really complex, and highly interconnected. It ignores the side effects that emerge when you put things together.

The tendency towards specialization helps to explain the failure of many expert predictions. The failure of "hedgehogs" to make correct predictions highlights the folly of trying to use simple models to deal with a complex world. Experts tend to be very narrowly focused. They are good at "linear" extrapolation within their own narrow field, but are often blind to the bigger picture. Almost inevitably, they fail to understand and anticipate issues and influences from outside their area of expertise. It was only in the latter part of the Twentieth century that science started breaking down the walls between silos of specialization. This is reflected by the recent trend to use compound names for new disciplines, such as bioinformatics, nuclear medicine, or ecophysiology.

For many people, their only familiarity with predictions lies in reading horoscopes. How is it that so many people report the success of fortune tellers at predicting their future? People who read horoscopes are looking for reassurance about the future. They reinterpret the wording in the context of their own lives. They ignore words that do not relate to them. Confirmation bias also plays a part. People ignore incorrect predictions, but every "correct" prediction serves to confirm their belief in horoscopes. Another reason lies in the nature of the predictions that fortune tellers make. If you make a prediction that is very specific then not only is it likely to be wrong, it will be obviously wrong. On the other hand if you couch a prediction in vague, general terms, then most readers will find something relevant in it. A prediction such as "it will start raining at 3 pm today" is so precise it is unlikely to prove correct, but a prediction like "it will rain" has a much greater chance of proving correct. Fortune tellers exploit the tendency of their readers to interpret everything they read or hear according to their own preoccupations and experience. Take this "prediction" for example:

Relationships are highlighted today. You leave behind any sense of regret as life warms up. Restrictions that tied you down no longer apply and you feel a sense of achievement.

The above statement is completely vague about details. The words are so general they are almost certain to apply to something. The word "relationships," for instance, could refer to any human interaction: love, family, work, neighbours. Likewise, "restrictions" could mean many things: a task you felt you had to complete, rain that kept you indoors or simply hesitation about spending money on a new dress. Whatever the case, readers will find aspects of their lives that fit the key words. They will latch on to phrases that match and ignore the misses. What is more, the horoscope is cast in such a positive tone that it is likely to become a self-fulfilling prophecy. A reader who wants it to come true is likely to highlight their relationships, cast off any regrets they many have, ignore restrictions and ultimately feel a sense of achievement from doing those things.

## What If?

In Chap. 6, we saw that problems can arise if planners use models that fail to take important issues into account. The trouble is that because the world is so richly interconnected, it is impossible to take everything into account. This means that the things we leave out can sometimes become just as important as the things we include.

The problem when trying to make predictions about complex systems is that they can be literally unpredictable. Even if a model captures the nature of the system perfectly, it can still behave chaotically. Chaotic systems are highly sensitive to initial conditions. Make a tiny change to a system here and it will become completely different over there.

If models are so unreliable, why use them at all? Why not experiment on the systems themselves? One reason is that there are some experiments we cannot do in real life. Some experiments are too risky, too big or too expensive. We cannot crash an airliner into crowded city suburbs just to test how well emergency services respond. We cannot release a deadly virus into the population to test whether we can contain it. But we can do virtual experiments.

Given that we cannot predict exactly what will happen, we do the next best thing and ask "what if" questions. What would happen if an airliner crashed? What would happen if a deadly virus were released into the population? We can simulate how a virus might spread or carry out exercises to test the response of emergency services as if a plane had crashed in the suburbs.

Usually the what-if questions we ask concern worst case scenarios. For example, we cannot predict when or where a forest fire will start, nor under what conditions. Instead we might ask what we would do to stop a fire that started in a dry forest on a day of high temperature and high winds and threatened to engulf a town that lay in its path.

A standard method when developing computer and engineering systems is to examine "use cases." That is, you ask how the system is to be used. You then test to make sure the system can handle every foreseeable case. This process includes what is known as "torture testing" in which software engineers test the ability of their design to handle extreme cases. For instance, if someone has not sent payment on their account, will the system still send them a threatening email even if what they owe is zero dollars? Of course, the platypus effect (Chap. 6) means that testing of this kind usually reveals flaws in a model. These flaws can then be removed by correcting the model.

The above approach to design and development of systems has proved extremely effective. A major event like the Olympic Games requires extremely complex organization, giving rise to endless possibilities for foul ups. At the Sydney Games of 2000, for instance, the organizers spent many months asking what-if questions to anticipate potential problems. Finding answers to these questions ensured that almost everything ran smoothly. They were so successful that at the closing ceremony, the IOF president declared the games the best ever.

Most questions in planning involve many different issues, all competing for attention. Problems arise when planners focus on one big issue and overlook the side effects of ignoring apparently minor issues. In the mid Twentieth century the idea of urban planning took hold in response to the rapid growth of cities. However, the planners were obsessed with the big issues of the time, especially the desire for impressive public spaces and the demand for roads to accommodate the growing numbers of vehicles. The effects on the way people live were often overlooked. Cities dominated by roads to cope with huge numbers of cars only encouraged more cars, which in turn created a demand for more roads. Alternatives such as walking, cycling, or public transport were minor considerations.

In most cities, cyclists are not allowed to ride on footpaths (sidewalks); they have to ride on the roads. This is a problem because modern roads and streets are designed for cars and make no allowance for bicycles. So the cyclists have to play dodgem with fast moving cars. In some cities this leads to hostility between the two groups: cyclists complain that cars drivers are rude and aggressive; drivers complain about having to swerve around bikes.

This problem often extends to pedestrians as well. My home street has no footpaths at all. So when I walk to the local shops, either I must walk across manicured lawns, and risk angering the neighbours; or else walk down the street and hope I can get out of the way when a car comes along.

Problems such as those described above certainly discourage anyone from trying to walk or cycle anywhere. So instead of walking, people climb into a car and drive a hundred yards to the local shops.[5]

My small problem with lack of anywhere to walk is symptomatic of a much wider problem in many cities, which are built around cars and not people. Modern urban designers, such as Danish architect Jan Gehl, are now working to change the approach of designers to urban landscapes.[6]

## The Art of Compromise

At Greed Incorporated (GI), only the bottom line matters. The investors expect profits and board members pride themselves on giving high returns. During the recent boom, GI's profits were unprecedented. The value of its stock soared. Its chairman hailed GI as one of the country's most successful companies. In the slump following the boom, however, its profits have stalled. The CEO argues with the board, insisting that the company needs to ensure its stability. But the board members are impatient for more profits. They fire the CEO and appoint a new one, who promises to make the company profitable again.

The new CEO believes in one thing and one thing only: the bottom line. He has no time for staff morale, no time for loyalty. He implements a policy of cutting costs and production efficiency, and demands strict economy from each division and section. This includes laying off staff. The survivors are grossly overworked. Resentment grows, leading to a high turnover of employees. Decline of corporate knowledge means that mistakes are made and opportunities missed. However the losses are well hidden because the CEO engineers takeovers of several smaller companies. He sheds their staff, strips their assets and takes over many of their accounts. Board members are seduced into compliance by chauffeur driven limousines and an opulent executive suite, but mostly by the handsome dividends they receive. After three years, the CEO has sold off all the company's viable assets to maintain profits. In a glow of achievement he retires with a huge golden handshake. Twelve months later GI is forced to file for bankruptcy.

Thinking and planning within a single model invites unexpected consequences. As the above story shows it sometimes leads to twisted strategies, such as the highly paid CEO selling off the company's long term survival for short term gains.

All too often elected politicians adopt the same sort of strategy, ignoring the country's long-term good for the sake of their own short-term popularity. Voters expect steady results from politics. The media judge government actions by weekly popularity polls. Their attitude is that if the approval rating falls even slightly, then the latest government decision cannot possible be correct. The only issue that matters is convincing voters to put them back in office at the next election. The government fears to adopt policies that will benefit their country in the long-term because that would risk losing votes and losing office. Instead, they implement short term measures that give the appearance of doing something and grab the credit if events elsewhere boost economic performance.

An underlying problem is that people try to obtain the best result within a narrow context. This is the problem of optimization. The issues are most easily understood in cases where the benefits can be measured. Take the case of the company in the story above. The goal for shareholders was to obtain the greatest possible profit. So the CEO sought to maximize short-term profit by increasing income and reducing costs, by any means. Within the narrow context of a single year's profit, his strategy worked. However, in the broader context, namely the company's long-term profits, his approach was disastrous.

When an individual considers a problem, he or she has a narrow point of view. Our understanding of the world around is limited by our individual knowledge and experience. This is a problem for decision making:

> ... our assumption that true intelligence resides only in individuals, so that finding the right person—the right consultant, the right CEO—will make all the difference.[7]

A second person looking at the same problem might see it differently and come up with a different solution. Each person brings a different perspective to a problem. This provides a strong argument for seeking diversity in groups.

> The positive case for diversity ... is that it expands a group's set of possible solutions and allows the group to conceptualize problems in novel ways.[8]

Problems arise when a group that makes decisions loses its diversity. This can happen if a single person makes decisions without consultation. Even when a group makes decisions, there can be problems. People who have worked together too long or have similar backgrounds tend to think alike.

## Dimensions of Being

Throw a stone into a pond and it creates ripples that spread outwards in all directions. The same thing happens in life. Take some action and its side effects ripple outwards, leaving behind a network of consequences. If these consequences are sometimes unexpected, it is because we look at the world through a narrow lens. We see only our immediate concerns. We fail to see all the other people, events or issues that may be affected.

Tracing associations makes a simple, but fascinating game. As a starting point, begin with a single idea or object, say a cat. In a circle around it, write down all the things it makes you think of (Fig. 11.1). Playing this game, you can very quickly put together a dense network of related terms. It is a network that is open ended: it could go on and on. Every link between terms represents some context, some model that reveals a link between two ideas. Think of a cat as an animal, and you

**Fig. 11.1** Some immediate associations flowing from the word "cat". Each word in this semantic map has a similarly rich number of connections. Note that every term in this network has other associations (not shown). They each form networks at least as rich as this one

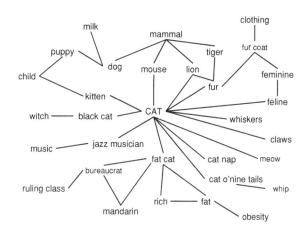

see relationships between cats and other animals. Think of "cat" as a slang term, and you get relationships to activities like jazz on the one hand, and to a host of similar slang terms on the other. Every different context you apply to the starting word produces new associations. For the word "cat," for instance, some associations might include "dog", "mouse", "mammal", "kitten", "milk", "witch", "black cat", "lion", "tiger", "whiskers", "jazz musician", "fat cat", "fur", "cat nap", "claws", "meow", "cat-o'-nine tails". The list could go on. Now look at any of the terms you come up with, "fat cat" for example. For this term, do the same as for the previous word. So associations for "fat cat" might include "bureaucrat," "mandarin," "rich," "ruling class," "obesity." Now take one of those terms. "Obesity" for example, might yield words like "weight," "lard," "health," "medicine," and so on.

The point of the above game is that there are many ways—many, many ways— to look at things. Nor is it a frivolous exercise; it has important uses and consequences, as we shall see. Previously (Chap. 6) we saw that context is a problem with intelligence testing. The testers assume that subjects *should* look at problems in a certain way, and regard other ways as wrong. In other words they regard children as being smarter if they think inside the square. But often, alternative ways of looking at things can be enormously creative.

A lot of humour arises from placing ideas in an unusual context, one that jogs our sense of the ridiculous.[9] Here is an example that plays on our sense of scale: Alice: *"What's worse than finding a worm in the apple you're eating?"*
George: *"Finding two worms."*
Alice: *"No, finding half a worm."*

We can create humour by looking at a situation from an unusual standpoint that makes things absurd. The reverse is also important. Looking at a situation from a new point of view can be a source of creativity.[9] Shakespeare's plays, for example, are packed with metaphors that invite the audience to see people and events in fresh, creative ways, such as in these famous lines from *Romeo and Juliet*:

> But soft! What light through yonder window breaks?
>   It is the East, and Juliet is the sun!
>   Arise, fair sun, and kill the envious moon,
>   ...
>   Two of the fairest stars in all the heaven,
>   Having some business, do entreat her eyes
>   To twinkle in their spheres till they return.

Something similar happens in science. A fresh perspective often leads to new insights. Serendipity, accidental events that lead to unexpected discoveries, is common in science.[10] It occurs most frequently when people do something new. For example, the Dutch scientist van Leeuwenhoek, built the first microscope and discovered single-celled organisms.

One of my former colleagues, an aquatic biologist, was prolific at making scientific discoveries. A specialist on the bugs that inhabited lakes, swamps and streams, he would use a microscope to look for bugs in water samples he collected during field trips. Almost every new lake he visited would reveal new kinds of bugs: new species, previously unknown to science.

His success reflects a well-known truth in scientific research. Try something new and serendipity will turn up discoveries for you. Going to new places has always been rewarding for biologists. It is no accident that the co-discoverers of natural selection, Charles Darwin and Alfred Wallace, both spent years travelling, observing and comparing living things in remote corners of the world. But more generally, new experiments, new technologies and new ideas all lead to discoveries. This is why, for hundreds of years, astronomers have been building bigger and better telescopes and each one yields important new discoveries about the universe. They may not know exactly what they will discover, but what they do know is they will discover something.

Physicists have been building bigger and bigger accelerators to smash atomic particles together to discover the fundamental constituents of matter. At the time of writing this, the world's biggest particle accelerator is the Large Hadron Collider (LHC). Built by CERN on the border of France and Switzerland, it is 27 km in circumference.

More subtly, looking at nature in new ways also leads to discoveries. In recent times, the idea of "natural computation," regarding natural processes as forms of computing, has led to many discoveries. Likewise, it has inspired computer scientists to look to nature, which has evolved many ways of solving complex problems. Reflecting this fresh perspective in science, entire fields of research today lie at the intersection of well-known fields and have double-barrelled names as a result: biophysics, molecular biology, biogeography, socio-economics, geographic information, bioinformatics, evolutionary computing, genetic algorithms, artificial intelligence, and artificial life.

The real world is every bit as richly connected as ideas. To take a familiar example, there are many ways in which people are connected to each other. To begin with, there is your family: your mother, father, grandparents, aunts, uncles, brothers and sisters, children, nieces and nephews, and of course your spouse. Next there are your neighbours, not just the people who live next door, or down the street, but everyone you have contact with in your community, such as shop keepers and city officials. Then there are the people you meet through work: colleagues, clients, suppliers, consultants and so on. Friends include people you know through clubs, societies, parties, music, and online social networks. And finally there are all the people you have contact with through miscellaneous means: the people we share the train with each day; the people who visit the same shopping centres, doctors or hairdressers as you do, the people you see at the gym, the pool, or the library. And each of these networks could be divided more finely as well. So we live our lives in the midst of many different social networks.

But social networks are not the end of things, not by a long shot. Many other kinds of networks impinge on our lives as well. There are transport networks, commercial networks, communication networks, and networks for essential services such as food supply, power, water, sewerage and so on.

All of these networks provide different contexts in which we live. Each of them is a potential source of huge complexity. And we live not just in one, but in all of them at once. But even greater complexity arises because every person, every item, in every one of these networks also exists in many different networks as well. The richness of these connections is central to our discussion. Many of the problems we have already looked at arise because every event we encounter is enmeshed in many different contexts.

Problems occur because the models that people bring to bear on a situation are never complete. Take the example of a road map we saw in Chap. 4. If shows the countryside as a scattering of towns, usually drawn as dots, connected by lines that indicate the roads between them. A map of this kind is perfectly adequate for finding the way to your destination. However, it leaves out many things. It leaves out hills and farms and forests. It leaves out cuttings and ditches. It leaves out current information such as road works, or local flooding. It leaves all the commercial and social life of the towns you may pass through on the way. It leaves out their history. In short, it leaves out almost everything except what is needed for the specific task in one specific context: driving from A to B.

In general, the problem is that models are limited, and what they leave out sometimes matters. It does not matter whether we are talking about schemas, the mental models we keep in our minds, or elaborate engineering models. The same limitations apply. No model includes everything.

This explains why diverse groups of people are often better at making decisions than any single individual.[11] Think of a decision as a network of possible issues. The real network will be rich with connections (e.g. Fig. 11.1). Any one person will see only some of those connections. In a diverse group of people, however, their different viewpoints and experience will cover a larger proportion of the network. This diversity helps to ensure that the group does not become stuck considering only a single context.

## The Platypus and the Echidna

The platypus and echidna are the only surviving members of an order of mammals called Monotremes. Throwbacks to our distant reptile origins, they are the only living examples of mammals that lay eggs. We can use these two archetypal Australian mammals to symbolize two kinds of problems we encounter when trying to make sense of complexity. Amongst other things, both problems are met by anyone who tries to organize information.

We already met the *Platypus Effect* in Chap. 6. It is a common problem encountered by designers in the course of developing information systems. It is closely related to another problem, which I will call the *Echidna Effect*. Trying to manage a complex system is like trying to catch hold of an echidna by its spines. Like the hedgehog or porcupine, the Echidna's body is covered in sharp spines. If you are careful enough, then each spine provides a way of grabbing hold of the animal. But it is much more likely, even if you are careful, that one or more spines will prick you when you try to catch hold.

The Echidna Effect is relevant to the process by which ants sort out their colony. As we saw in Chap. 2, ants build a colony by sorting objects in *ad hoc* fashion. The same idea has been applied in computing to produce the ant sort algorithm, a method of sorting borrowed from nature.

When we try to put the idea into practice, however, the Echidna Effect immediately presents thorny problems. Suppose that the objects we want to sort are documents on the Internet. At first sight it seems simple. We can use virtual ants to sort the documents into lists of similar documents. But it is not quite that easy. How do you classify the documents? What do you base the categories on? Words used in the title? Name of the author? Publisher and date? Key words? Or words in the text itself? As if that were not bad enough, you then have problems of generality. For instance, a novel is a special kind of book, and a book is a special kind of publication. So how do you deal with these relationships? It is a thorny problem. Perhaps it is appropriate that Echidnas eat ants.

The Echidna effect leads to another important phenomenon: cascading contexts.

## Cascading Contexts

A car owner is short of money, so he puts off getting his car serviced. Lack of maintenance allows a problem to develop in the car's brakes. In turn the brake problem means that a few days later he is unable to stop in time when another car suddenly pulls out from a side street.

When we seek simple solutions to problems, issues can arise from other contexts. Sometimes these problems build on one another. Each problem sets off further problems in other contexts. Like water gathering speed down a waterfall they create a cascade of problems. The end result can be a problem far greater than the one we first tried to solve. As the poem at the start of Chap. 9 showed, even ignoring a missing nail in a horseshoe could lead to losing a war.

Cascades like the above occur because the real world is more complex than any model people might use in their thinking or planning. Events in one context can trigger conditions in another context, setting off a whole new chain of events. Such problems defeat conventional methods for interpreting model, such as scenarios and sensitivity analysis, because they arise from conditions outside the model concerned.

Cascading contexts pose a constant problem for decision-making. The problem in asking "what-if" questions is that you need to identify all possible combinations of circumstances that might occur. But given the complexity of the real world, this can be virtually impossible.

Product manufacturers face this problem all the time. They can anticipate many problems, but the real world always throws up situations they never thought of. In later chapters we will look at the ways in which cascading contexts have led to unexpected side effects of new technologies.

In engineering, the problem of unforeseen contexts forces engineers to see technical development as a cycle. Development of a major technical system is never regarded as complete. You always expect the unexpected. So developers keep track of how a system performs and when problems arise, they tweak the design to allow for the new circumstances. This cycle of implementation and monitoring goes on and on.

As we saw in Chap. 9, trying to exercise due care and eliminate potential safety problems can also become an endless cycle of applying band aid solutions to cover up gaps. In some activities, such as air travel, the concern over safety problems is so great it has prompted international efforts to address the matter.

Prior to World War 2, air travel was a luxury enjoyed only by the rich and powerful. The post-war years saw a rapid expansion of public air travel around the world. Accompanying the increase in public air travel was a steady increase in the annual number of deaths resulting from air crashes (Fig. 11.2). Two factors were at work. One was the increasing number of flights; the other was an increase in the size of aircraft. In other words, more and more crashes occurred each year, and each crash killed more and more people.

This upward trend reached its zenith in the 1970s. The worst year of all was 1972, when there were 2,374 casualties from airline accidents around the world.

The horrific nature of crashes in which hundreds of people died made sensational news reports. Graphic pictures of people dying in gigantic fireballs are enough to scare anyone. Airlines and aircraft manufacturers are naturally sensitive about the impression that tragic incidents burn into the minds of the public. The resulting threat to the aviation industry provided enormous incentive to remedy the situation.

For decades the airline industry has worked hard to make air travel as safe as humanly possible. Countries with the greatest vested interest set up bodies to enforce transport safety (this was not confined to air travel, but also included travel by road, rail and water). Agencies such as the International Air Transport

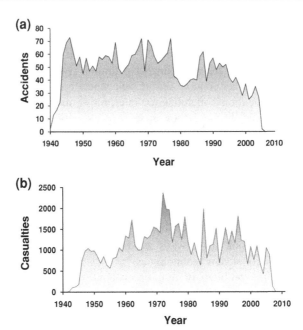

**Fig. 11.2** Trends in airline accidents over time. **a** Number of accidents per year world-wide. The average number was constant until the 1980s, then declined as regulations improved safety. **b** Number of casualties in airline accidents per year. Note the decrease in both accidents and casualties since the late 1970s.[13] There was an upward trend in casualties until the mid 1970s as the seating capacity of aircraft increased. Casualties then declined as safety measures reduced the frequency of accidents

Association (IATA) and International Civil Aviation Organization (ICAO), as well as numerous national authorities, investigate every air accident. The Aircraft Crashes Record Office (ACRO), based in Geneva, records details of every accident.

Air safety and security have been matters of closing the barn door after the horse has bolted. Aircraft operations are so complex, that it impossible to anticipate every problem that might arise during flight. But every time a plane crashes, teams of experts undertake a detailed investigation to try to discover exactly what led to the crash. In effect each accident became an experiment that provided vital information. The results helped to frame procedures and certification requirements for aircraft design, maintenance procedures, pilot training and air traffic control. In short, each death has helped to save lives in the future.

Over the years, this work has led to international standards for aircraft manufacture and maintenance, as well as procedures. The result has been that air travel has steadily become safer over the years.[12]

From the late 1970s onwards there has been a steady decline in the annual number of airline fatalities worldwide (Fig. 11.2). And this decrease has been achieved in spite of continued growth in the size of aircraft and numbers of passengers. Statistically, flying is now a safe way to travel.[13]

---

## End Notes

[1]Paul Valery In *Analects,* vol 14 of Collected works ed. J Matthews 1970 Routledge and Kegan Paul, London.

[2]Tetlock (2005).

[3]Gardner (2010).

[4]Edward Teller, quoted by Robert Coughlan in an article titled "Dr Teller's Magnificent Obsession" in *LIFE* September 6, 1954. p. 74 (pp. 60–74)

[5]In his book *A Walk in the Woods*, Bill Bryson vividly described this same problem in his own home town.

[6]Jan Gehl (2010).

[7]Surowiecki (2004) p. 36.

[8]Ibid.

[9]Koestler (196?).

[10]Green (2004).

[11]See an earlier discussion in the section The Art of Compromise. Also see Surowiecki (2004).

[12]Source http://aviation-safety.net/.

[13]See Chap. 7, End note 11. Also see http://www.nsc.org/news_resources/Resources/res_stats_services/Pages/FrequentlyAskedQuestions.aspx.

# Collateral Damage

12

*And what shoulder, and what art, Could twist the sinews of thy heart? And when thy heart began to beat, What dread hand? and what dread feet?*[1]

**Abstract**

Every new technology has social side effects. Most are unintentional. Massive social changes accompanied the Industrial Revolution. The Information Revolution is likely to set off equally wide-sweeping changes in society. Notable changes are the appearance of new media for communication, the use of "big data" in decision-making and the spread of automation. Rapid increases in the numbers of new patents imply that technology is changing faster than ever. Numerous trends imply that western societies are changing very rapidly as a result. Side effects of automation, for instance, have contributed to the loss of many jobs and to increasing social isolation.

## New Technology

When Europeans first visited Shaka, king of the Zulus, in the early Nineteenth Century, they showed him examples of modern technology.[2] On seeing samples of writing, the warrior king is said to have expressed no interest at all. His men, he said, were trained to relay messages accurately on pain of death.

When people are confronted with new technology they rarely understand its potential, let alone see its wider implications and long-term impact. What was true in King Shaka's time is still true today. In the latter part of the Twentieth Century, advances in communications technology led to many cultural changes. Within the space a few decades, we went from a printed culture to an electronic culture and

D. G. Green, *Of Ants and Men*, DOI: 10.1007/978-3-642-55230-4_12,
© Springer-Verlag Berlin Heidelberg 2014

finally to an Internet culture. Facing such rapid change, many people have found it hard to adapt.

In the early 1990s, I spent much time demonstrating and explaining a strange new technology to professional groups in business, government and academia. Few could see any use for it. Most thought it a pointless waste of time and money. The new technology was the World Wide Web. A typical reaction amongst business people was to dismiss the new technology out of hand. Why waste time with a technology that only academics used? None of their customers would see it. Their current systems for record keeping and advertising were more than adequate. Even academics were hard to convince. When I opened my LIFE website in 1993, the director of my institute thought it was a disgrace. He said it would spoil the institute's reputation. It apparently escaped him that attracting 10,000 online visitors each day to read about the institute's research was actually enhancing the institute's reputation.

Twenty years later, those early hostile attitudes of many business people seem short-sighted. Today, distributed information is a crucial tool in research and business. Everyone expects to be able to learn about a company's products and services via the Web, and vast business empires have grown up around the technology.

An experiment I carried out in 1993 revealed the Web's true potential. After presenting a talk at a local conference I wanted to make my results more widely available. So I placed a copy of the article on my web site for colleagues to download. A month later, I was astonished to find that more than 10,000 copies had been downloaded.

## Collateral Damage

Developed and introduced to help solve one particular problem, new technologies often have unexpected side effects. All too often, these side effects create as many problems as the new technology solved. In some cases, they even make the very problem they were intended to solve worse than before. Tenner called this the "Revenge Effect."[3] The introduction of computers, for instance, was supposedly going to lead to the "paperless office". Instead it had the opposite effect.

Computers have solved many problems, but they also introduced many others. They gave us the ability to process data faster and in greater quantities than ever before; but they also gave us computer viruses. They enabled us to exchange data with people all over the world in the blink of an eye; but they also gave us spyware.

In industry and commerce, automation has led to huge increases in efficiency. However, many of the innovations have unfortunate side effects. One of these is the loss of jobs for workers whose tasks are taken over by machines. In an old-fashioned grocery store, for instance, each customer would hand over a list of items and wait while the store-keeper went to collect them. Meanwhile, the customers would queue up waiting. It is far more efficient to let the customers collect their own goods from supermarket shelves. Likewise, petrol station owners cut the cost of hiring attendants by providing the means for drivers to operate petrol bowsers themselves.

Automated teller machines enable people to withdraw money anywhere and at any time. They also reduce the cost of hiring tellers to handle account transactions. Paying bills by phone, Internet banking and credit card transactions eliminate the need to go near a bank at all.

The above innovations also enable banks to make huge savings. They ensure that the companies concerned are paid on time and the banks profit from the service fees. For customers, credit cards and automated bill payments free them from having to travel all over town making cash payments. However, one side effect of using credit cards is that card owners no longer hand over cash, so they can easily fail to notice how much they spend. What is more, they can even spend money they do not yet have. It is all too easy for them to spend large sums without realizing it. Credit card debt is now a major problem in many countries. In the United States alone, credit card debt currently exceeds $800 billion.[4]

Automation also contributes to personal isolation. When doing business over the phone at least you got to talk to another human being. But even that small amount of contact is being eroded. Most large companies now use online menus, voice recognition and speech generation to automate their telephone services. Now you talk to a real person only as a last resort.

A paradoxical effect of advances in communications technology and automation has been to make people more isolated from one another. Online banking is just one example where people no longer need to make human contact when conducting their personal or business affairs. Just consider entertainment for a moment. In the mid Twentieth Century, one of the main venues of entertainment was the theatre, either live or cinema. During a single lifetime, an entire culture—the drive-in cinema—grew up, thrived in western countries for a time and eventually died. In the late 1950s my family revelled in the novelty of watching a movie from our own car. Before long, my parents were taking us kids to see a drive-in movie almost every week. However, the introduction of television changed all that. Why pay money when you can sit at home and see movies for free? By the year 2000, every drive-in theatre I ever visited had been closed down.

The social impact of television is that people no longer need to leave the house for entertainment. The introduction of videos, DVD and home theatres make staying at home even more attractive. With videos, people at least had to visit a video library to select a movie; with online movies on demand they just click a button on their desktop and it is delivered to them. And the trend away from face-to-face social contact just keeps on increasing. In later chapters we will look again at this problem of increasing social isolation and its consequences.

## Benri Desu Ka?

A common question you get asked while travelling in Japan is "Benri desu ka?" which means "Is it convenient?" So you are regularly asked very politely whether your hotel room is *benri* (convenient), whether the travel arrangements are *benri*, whether dining now is *benri*. So much of modern society, especially in service industries, is about making life more convenient for people.

Human societies first emerged as a survival strategy, as a way of ensuring that the necessities for survival would be met. In modern society, however, the machinery of survival normally runs so smoothly that people can afford the luxury of seeking convenience. That is, they are able to spend less time and effort on things they need to do and more time on things they want to do.

Necessity may be the mother of invention, but in a fiercely competitive commercial world, necessity often morphs into convenience. Many modern inventions are about making life more convenient for people. Washing the dishes used to take up a large part of every evening, but popping plates into a dishwasher is so much more convenient. Likewise, shopping in a supermarket is more convenient than spending half a day wandering from one shop to another tracking down the things you need. Shopping online is even more convenient still. Watching a movie on video is more convenient than coping with traffic and crowds when going to the cinema.

Consumer convenience is a marketer's necessity. For a busy person, it is more convenient to buy food, whether at a restaurant or a takeaway, then to spend hours preparing meals at home. Meeting this need creates a marketing opportunity.

As the above examples suggest, convenience often equates to saving time, saving effort, or making things simpler. Most telling is that, by saving effort, innovations create opportunities for people to do more, or to do different things. And these opportunities create side effects that can lead to social change. In later chapters, we will look in detail at the impact that the introduction of labour saving devices had on family life and on the ways people live during the late Twentieth Century (Chaps. 15–17) as well as the impact of new communications technology on the patterns of social connections that people form (Chap. 14).

## Technological Change and Society

One of the most important long-term trends in human history, especially in the last 300 years or so, is the accelerating pace of technological change and its impacts on human society.

Sometimes the side effects of a new technology can shape an entire society or culture. A natural experiment from historic times highlights the way new technology can reshape a society.[5] In the Seventeenth Century, two groups of Puritans left Britain to settle in the New World. They aimed to create societies modelled around their religious beliefs. One group settled at Plymouth in New England. This group subsisted for years on crops of maize. A second group settled further south, on Providence Island off the coast of Nicaragua. Here the settlers soon discovered that conditions on the island were ideal for growing sugar, a crop that is both easy to grow and highly profitable for trade. The nature of sugar cane farming, which requires much less care and attention than grain crops, made it possible to increase profits by exploiting slave labour. The result was that the settlers abandoned their Puritan ideals and became slave owners.

Historically, several great transitions in society are associated with new technologies. The development of farming and domestication of animals, for instance, provided a steady food supply, which was an essential step in the transition from nomadic hunting and gathering to the formation of large settled communities. Such communities then set the scene for the advent of civilization, which is associated with several new technologies, especially writing, and building in stone. During the Middle Ages in Europe, the invention of printing led to the growth of modern science, as well as sowing the seeds for new kinds of social order.[6]

More recently, the Industrial Revolution was accompanied by waves of social change. These included an increasingly urbanized population; the growth of cities; and social revolution, especially moves to democracy and other new kinds of government. All of these changes were accompanied by violence and enormous social upheaval, including the French and Russian Revolutions, the Napoleonic wars and the two world wars of the Twentieth Century.

Numerous studies have traced ways in which particular technologies helped to transform society. These historical changes show that new technologies have set off cascades of social change in the past.

One example, from the Industrial Revolution, is the series of inventions that mechanized the textile industry in Britain. By the early 1700s, home weaving was a widespread cottage industry. It was based on two traditional tools: the spinning wheel, which made thread, and the hand loom, which weaved the thread into fabric. Over a period of several decades a series of inventions made it possible to automate the entire process from raw wool or cotton to finished fabric. These inventions opened the way for the industrial scale production of textiles.

The changes started in 1733, when John Kay invented the flying shuttle.[7] This allowed pairs of weavers to be more productive than both working alone in the traditional way. However, production was still limited by the rate at which cotton could be produced. The automation of spun cotton was achieved by a series of inventions. In 1764, James Hargreaves invented the Spinning Jenny.[8] Subsequent refinements mechanized the process as well as making it more reliable and efficient. By the 1780s, cotton mills were able to undertake cotton production on a large scale. Other inventions, made it possible to automate weaving. By the early 1800s, weavers were using sequences of punched wooden cards (a forerunner of today's computer programs) to enable looms to produce complex fabric patterns.

The social effects of the Industrial Revolution were mixed. In the long term it eventually led to greater prosperity in the countries concerned. But in the short term, it triggered a cascade of social problems that lasted for many decades. Displaced farmers and cottage workers flocked to the cities, creating large-scale squalor in which poverty, crime and disease were the norm. While moral reformers succeeded in having slavery abolished, millions of factory workers became virtual slaves in collieries and "dark satanic mills",[9] where they laboured long hours in dreadful conditions for little pay.[10]

In Britain, misinterpreting the masses of unemployed poor as lazy led to the introduction of poor laws and the formation of poorhouses, essentially prisons for the crime of being poor. At the same time, high crime rates, combined with harsh penalties, led to overflowing prisons. One solution was transportation, which helped to spread English language and culture, first to North America, and later to new colonies in Australia.

Mechanization meant that many tasks required little knowledge or skill. Seeking to minimize labour costs, managers hired the cheapest workers they could find. With no legal impediments managers were free to hire young children, work them for long hours in dreadful conditions and pay them wages that were barely enough to live on. Child labour became the norm in mechanized industries.

Disaffection of the poor inevitably led to political changes. These included the French Revolution, the origins of Marxist ideas and communism and later, the Russian Revolution.

## Economies of Scale

Positive feedback is a surprisingly common mechanism in unplanned social change. It occurs whenever the result of a change is to produce more change of the same kind. As we saw in Chap. 10, positive feedback makes things grow, and not just grow, but grow at an ever increasing rate. This leads to exponential growth. Several effects follow. One is that small, local features can suddenly grow into global patterns. If unchecked, a disease introduced by a single infected individual can rapidly spread to engulf most of the population of a city or country.

There are many other cases where positive feedback leads to exponential distributions. It often comes into play when an activity can grow indefinitely. This frequently happens in socio-economic settings when features can be standardized. This allows them to become scalable: they can be reproduced endlessly, in any number of places.

Take marketing chains, for example. Each chain has a standard business model, which can be replicated anywhere. New franchises do not need to formulate a business plan; they just apply the standard model that is given to them. They do not need to find equipment or organize advertising: the chain supplies it all. The bigger the chain becomes, the faster it can grow. Supervision of new franchises can be delegated to regional managers.

Some activities are scalable; others not.[11] For example, up until the Twentieth Century, acting and theatre were activities that were not scalable. If you wanted to see a play, then you needed to attend a live performance in person. Making this possible provided a role for theatres and actors in any town of any reasonable size.

However, the advent, of entertainment technology changed all that. First came cinema, then television, video tapes, DVD, and downloads. The upshot was that a single performance could be endlessly repeated anywhere, anytime. Why watch a performance by second rate local performers when you can watch award winning performances by the world's best whenever you want?

The same principle applies to music, literature, news reporting and other information related activities. The result has been revolutionary, not to say devastating for many. Scalability means that positive feedback comes into play. Success breeds success. Instead of being mere local celebrities, some actors become world famous and unbelievably rich. On the other hand, instead of enjoying a viable, if modest, career in a local company, most aspiring actors have to struggle to get any parts at all. The situation for writers is similar. Go into a bookstore and you find the shelves dominated by a small number of mega-popular authors and titles. Many brilliant authors never even succeed in getting their work published at all.

Since the turn of the Millennium, Internet technology has added new twists to this story. The ability to download books, music and videos has put pressure on more traditional, local outlets. When you had to obtain physical copies of books, say, then there was a role for local bookstores. But the ability to download e-books has reduced sales for bookstores drastically. At the time of writing, several well-known chains of bookstores have gone out of business. The same is not true of libraries, which survived by reinventing themselves as information resources.

Of course, old institutions have not disappeared entirely. Live theatre retains the immediacy of social interaction. Movie theatres have survived successive onslaughts from TV and video by increasing efficiency and upping the ante with larger-than-life screens and 3D.

New technologies are also changing the career path for some professions. For decades, aspiring singers were at the mercy of large record companies. If they agreed to record your song, you were made; otherwise you were nowhere. Now however, singers can bypass this system entirely and simply release their own performances online. If your upload "goes viral" then you are instantly world famous. Similar options now exist for aspiring writers. Mind you, it still applies that many try but few succeed.

At the time of writing this, other industries are suddenly becoming scalable as the effects of new technology make themselves felt. One of these is the retail industry. Although it is only information products that you can download directly, communications are creating new alternatives to the traditional model of visiting the local shops. The ability to order online and have products delivered saves time and effort for individuals. This is especially true for products that are hard to find. Any retail outlet has only finite space to house and display products. The result is that shoppers often need to go from store to store trying to find the exact product they are looking for. It is much easier to search for products online and have them delivered.

While online ordering has helped large chain stores to increase their dominance in the marketplace, it has also helped retailers dealing in specialized products. Formerly, sales for such stores were limited to people who could make a special trip to visit them. Now their potential marketplace is worldwide. Online trading also enables manufacturers to bypass retailers and sell directly to their customers. One of my favourite wineries, for instance, is small family business operating just outside the town of Rutherglen in northern Victoria. No store that I know of sells their wines, so for years my only way of obtaining bottles was to make a pilgrimage to the winery itself. Now they have a website where I can order the latest vintage and have bottles delivered direct to my door.

Of course, going shopping is a traditional social activity, which many people enjoy. One response is for stores to become advisory centres that hold certain popular products in house, but act as portals where buyers can obtain expert advice about a much larger range of products.

The information revolution is poised to affect many other industries and professions. One success of artificial intelligence has been to learn how to encapsulate bodies of knowledge from specialized fields. Knowledge is essentially about using information to do things. It can be expressed as rules, such as those we met back in Chap. 2. This means that a set of rules can often capture the essential knowledge of an expert in a particular area. These *expert systems* summarize what professionals know so that others can apply their knowledge to solve practical problems.

Within present technology, visits to the doctor are not scalable. For any serious condition, a doctor needs to see and examine you in person. Wisely or not, however, people can now exploit online resources to diagnose and treat themselves for simple complaints.

Similarly, online services are appearing to assist people with financial and legal matters. The law has always been an arcane world replete with rules and procedures. However, some legal procedures, such as writing a will, are relatively straightforward. Online kits are already widely available that provide users with step by step guide through some of these simple legal procedures.

The same is happening in some areas of accounting and finance. In particular, it is in the interest of governments for citizens to pay their taxes as quickly as possible. Some governments therefore provide online programs that guide taxpayers through the steps involved in preparing their tax return.

The local stock market used to be an exciting place. It was crowded with buyers and sellers shouting at each other while recorders used chalk to write up the latest prices on a massive trading board. Only registered stockbrokers were allowed to buy and sell, so if you wanted to trade in shares you would ask one of them to make the trade for you. Today anyone can buy and sell shares online.

Stockbrokers and financial institutions use software "agents" to do much of their buying and selling for them. In effect you get the situation where one computer program is selling shares to another program. This *algorithmic trading* has changed the way the share market behaves. To take a simple example, suppose that an agent is programmed to buy whenever it detected an upward trend in a share price and to sell when it detects a downward trend. Then it can take advantage of rapid fluctuations in share prices to get the best deals for its owners. What is more, it can do this much faster than a human could.

The problem is that because everyone uses these "intelligent agents," unexpected effects emerge. For example as soon as an upwards trend becomes apparent, then software agents all over the place start madly buying up shares. And the faster they buy, the more rapid the trend. This creates a positive feedback loop and before you know it, share prices have boomed. Likewise, a slight downward trend can lead to a selling frenzy and prices crash spectacularly. Of course, boom and bust cycles have always been common in the share market. However, automated trading has led to market fluctuations that are much faster and sharper than ever before.

The programs used by traders today are highly sophisticated. They can respond to a great variety of information from many different sources. They can even scan news reports and company releases for clues about coming trends. In high-frequency trading, being even a few seconds ahead of a rival can mean the difference between profit and loss.

## Serendipity

In 1609, Galileo, the famous astronomer, read about a wondrous new device, the telescope, invented in Holland by Hans Lippershey. Within a few months Galileo had acquired lenses and made his own telescope. Being an astronomer he immediately pointed his new toy at the night sky. On 7 January 1610, he made a startling, and unexpected discovery. Directing his telescope at Jupiter, he found what appeared to be four small stars nearby. Over the winter he continued to observe these objects and discovered that they were actually satellites revolving around Jupiter. In the Seventeenth Century this discovery was not only unexpected; it was revolutionary, and dangerous. Religious doctrine at the time held that Earth was the centre of the universe. It held that all heavenly bodies revolved around the Earth. So to find satellites revolving about Jupiter refuted Church doctrine. That was heresy. But once it was known, the discovery helped to reinforce the heliocentric theory of the universe.[12]

Serendipity, unintentional discovery, has played a key role in science since its very beginnings. As we saw in the previous chapter, new ways of looking at things usually lead to discoveries. For example, a side effect of European settlement of Australia in 1788 was the discovery of the platypus. In Chap. 6, we saw that this unexpected discovery by the early settlers forced scientists to rethink their model of the animal kingdom.

Serendipity occurs when new connections are made. Koestler argued that creation is the act of seeing everyday things in a new light. Look at a wine press, see a connection with printing, and you have the printing press.[13] Do something different and you are bound to discover something new. You may not know what you will discover, but you know that you will discover something. Sail west, like Columbus and you might discover a new continent. Go to Australia and you might discover a new animal. Go to a pond no one has studied before and you are likely to find new kinds of bugs living there.

Like Galileo and his telescope, new technologies have always led to serendipity in science. The first microscope led to the discovery of microbes; X-ray crystallography led to the discovery of DNA's helical structure. It is no wonder then that physicists always want to build new kinds of telescopes and bigger particle accelerators, as we saw in the previous chapter.

Not surprisingly, computers have made many scientific discoveries possible. In the book *The Serendipity Machine*, I described how information technology is both a source of serendipity and a tool that is changing the way science is done. The main effect is the explosive growth in data: up until the mid-Twentieth Century data was rare and expensive; today it is abundant and cheap. The volumes of data collected in almost every sphere of activity are now so great that the term *Big Data* was coined to refer to the phenomenon and to uses that can be made of it.

Information technology makes it possible to gather far more data than ever before. Medical instruments continuously record vital signs of patients. Radio devices enable biologists to track the wanderings of birds and animals in the wild. Automatic weather stations collect data from thousands of locations around the globe. Satellites monitor the entire Earth's surface on a regular basis.

The Internet makes it possible to combine and share data. The Internet itself grew out of the need for defence contractors to share resources. Today almost every field of research has specialist repositories where researchers share data, information and software. In biotechnology, for instance, data about genes, proteins and other materials are stored in three vast international repositories: the DNA Database of Japan (DDBJ), the European Molecular Biology Laboratory (EMBL) and Genbank.

The sheer volume of scientific data now available makes it possible to interpret data in new ways. In the early 1990s, I was involved in a project to provide the government with comprehensive information about Australia's environment. As we compiled all the necessary data, an interesting phenomenon emerged. Not only could we use the data to answer all the questions we needed to answer, but also we could answer many other questions that no one had thought of before. It soon became apparent that with all this data linked together for the first time, almost anything you cared to do with the data would lead to new and potentially significant discoveries. We referred to this phenomenon as the *Serendipity Effect*.[14]

This Serendipity Effect is today the basis for an entire field of research, known as *Data Mining*. By combining different kinds of data, you can discover all kinds of useful things. Combining, say, phone records and credit card statements can reveal interesting patterns that allow companies to target their advertising more precisely. Tax departments, for instance, use data mining to identify potential tax evaders. Data mining is an important tool dealing with Big Data.

As shown earlier by the example of the printing press, new ways of thinking, and new ways of looking at things inevitably lead to new discoveries, to serendipity. Every age sees the world in terms of its own preoccupations. During the industrial revolution, scientists tended to treat the world as a kind of machine: in the human body for instance, the heart was a pump, the digestive systems was a power supply and the brain was a control system. As the Information Revolution

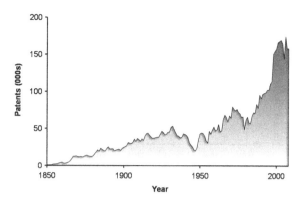

**Fig. 12.1** Trend in the number of patents processed each year by the United States Patent Office. As the figure shows, the rate of new patent approvals per year grew exponentially during the second half of the Twentieth Century. *Source* USPTO

gathers pace, scientists are increasingly interpreting the world as "natural computation." This idea has led to new models of living systems. It also sees scientists solving complex problems in computing by borrowing ideas from nature. Some areas of advanced computing today are almost indistinguishable from biology. Artificial neural networks (ANNs) learn solutions to problems by creating networks of "neurons" that process data and outputs solutions. Evolutionary computing uses serendipity deliberately. One of the most widely used examples is the genetic algorithm. Given a complex problem it evolves solutions. It does this by maintaining a population of potential solutions and repeatedly "breeds", each time selecting the best ones to keep and breed again.

In Chap. 2 we saw how ants build an ant colony by following simple rules of behaviour. Used in computing this process is known as the ant sort algorithm. It can be adapted to sort items on an *ad hoc* basis. Other aspects of ant behaviour have also been borrowed for computing. Ants travel to food sources by following pheromone trails left by foraging ants. As more and more ants follow the trail they gradually smooth out the bends and twists, making the trail more direct and shorter. This idea lends itself to a computer algorithm known as *ant colony optimization*.

## Torrents of Change

Sitting right at the centre of American economic innovation is the United States Patent and Trademark Office. Opened in 1836, the patent office was established to record details of new inventions and to act as umpire in cases involving disputed ownership of profitable technologies.

Since its founding, the Patent Office has processed new inventions in ever-increasing numbers. Checked only by world-shaking events, such as World War II, the number of new patents has increased steadily (Fig. 12.1). The boom in information technology at the turn of the Millennium contributed to the rate doubling in just a 10 year period. At the time of writing, the US Patent Office processes more than 200,000 patent applications every year. It employs an army of

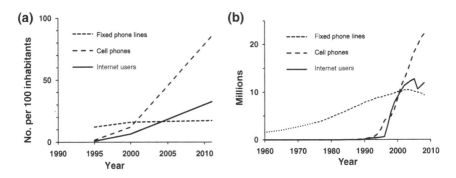

**Fig. 12.2** Over the period 1995 to 2005, a phase change took place in the way people communicate. Prior to 1995, most communication was by *fixed phone line* (*dotted lines*). By 2005, most communication was via mobile/cell phones (*dashed lines*) and the Internet (*solid lines*). **a** World-wide trends. **b** Trends for Australian users

5,000 patent experts to keep track of them all. By 2010, it had recorded over 7.5 million patents in all: more than two thirds of them in the last 50 years.

Enormous though the work of the US Patent Office is, it is not the busiest such office. In the period 2003–2007 the Japanese Patent Office processed some 2.35 million patents, as against the 1.66 million patents by the US Office.

One of the most important long-term patterns in human history, especially in the last 200 years or so, is the accelerating pace of technological change. In the U.S. the rate at which new patents were lodged per annum increased by nearly an order of magnitude between 1900 and 2000. And in the space of just 10 years from 2000 to 2010, the rate doubled again (Fig. 12.1).

We see some of the immediate effects of the rapid turnover of patents in information technology. Technological change is now so rapid that we have to update software and hardware on a regular basis, or risk being left behind by industrial norms. Another example is the change in the way people view motion pictures. Cinemas first became popular in the early 1900s. After the introduction of sound in the late 1920s, cinema rapidly overtook stage plays as a popular form of entertainment. Television was introduced in the 1930s, but only took off as a mass medium in the 1950s. Movies on videotape became popular through "video libraries" in the 1980s. They were completely replaced by DVDs in the early years of the new Millennium. By 2010, Internet downloads were becoming increasingly popular. As this sequence shows, the technology governing the context in which we see movies has changed at an ever-increasing rate.

Another major transition can be seen in communications. During the period 1995–2005, people in most developed countries experienced a phase change in the way they communicate (Fig. 12.2).[15]

All of this means that we are currently experiencing technological innovation at a greater rate than ever before in history. It is so great that technological change is not the exception, but the norm. Given that new technology leads to social change, then this torrent of new technology suggests that we are likely to experience social upheavals faster and greater than ever before.

What evidence is there to indicate that we are indeed experiencing substantial social change? One kind of indicator for social change lies in social trends. One is the pace of modern life. Numerous authors have pointed out that society is speeding up.[16] The world at the start of the Third Millennium is like the Industrial Revolution on speed.

An important characteristic of complex systems is that positive feedback can turn small local variations into global patterns. In 2007, the pollster and social analyst Mark Penn identified seventy-five "microtrends" currently taking place in American society.[17] The trends he identified included changes in family life, leisure, politics, and religion that are transforming people's lives. Amongst other things, he pointed out that small patterns of behaviour can spark large-scale changes in society as a whole:

> Only one percent of the public, or 3 million people, is enough to launch a business or social movement.

He goes on to point out that

> Entrepreneurs are using the Internet, viral marketing, and social networking to identify and profitably promote their products and services in ways not possible only a few years ago.

Microtrends such as these are just the tip of the iceberg. In later chapters we will examine several trends in more detail. These include the effects of new technologies on family life, on society, and on the world's environment.

---

## End Notes

[1]Verse 3, from the poem *The Tyger* by William Blake.
[2]Morris (1965).
[3]Tenner (1996).
[4]U.S. Census (2012). The 2012 Statistical Abstract. *The National Data Book* www.census.gov.
[5]Bowles (2011a) provides a detailed investigation of the Puritan "experiment". Bowles (2011b) provides a popularized overview.
[6]Fang (1997).
[7]Deane (1965).
[8]Ibid.
[9]From the poem Jerusalem by William Blake (1808). It was first published in the preface to Blake's *Milton, a Poem*. It became a popular hymn when Sir Hubert Parry set it to music in 1916.
[10]A modern parallel would be sweatshops in Third World countries that produce cheap goods for Western markets.
[11]Taleb (2007), Chap. 3.
[12]Koestler (1964).

[13]Ibid.

[14]Ibid.

[15]World figures are based on United Nations (2013); Australian figures are based on ABS (2009).

[16]For example, Gleick (1999), Rudel and Hooper (2005), Posente (2008).

[17]Penn and Zalesne (2007). Quoted from a commentary by the authors 14 December 2007 in *The Millionaire Mind Support Network*. http://mymillionairemind.org/author/sandi/page/13.

# The Straw that Broke the Camel's Back

## 13

*When anything is used to its full potential, it will break.*[1]

**Abstract**

The network model reveals deep similarities between social changes and many natural phenomena. The spread of ideas and epidemics, for instance, are both examples of percolation. An important process is the "connectivity avalanche" in which a network undergoes a phase transition from a fragmented set of nodes to a single connected whole. Models of complex systems rely on systems being closed, disconnected from events out of the norm. Extreme events can produce a cascade that connects normal events to a wider network in which many new and unexpected events become possible. Among other things, this change helps to explain the unexpected way in which major accidents occur.

## Why Is a Starfish Like an Atomic Bomb?

Some of the greatest discoveries in science have come about by seeing deep similarities, especially between things that at first glance seem unrelated. How, for instance, could a starfish share anything at all in common with an atomic explosion? They are completely different. And yet, if we look closely they are similar. The secret is to look at them as networks.

To begin, we can think of an atomic explosion as a network. The explosion starts when a neutron hits the nucleus of a uranium or plutonium atom. As the nucleus splits, it releases two more neutrons. These neutrons split two other nuclei, which in turn release two more neutrons each. And so the chain reaction spreads. This is a network in which the atoms are the nodes and the spreading neutrons make the edges that link different atoms.

So what has a nuclear chain reaction got to do with starfish? The answer stems from the nature of coral reefs. Take the case of Australia's Great Barrier Reef. It is not a single reef at all. It actually consists of thousands of separate coral reefs scattered along the east coast of Queensland. In the mid 1980s there was widespread alarm in both the tourist industry and conservation agencies.[2] Crown-of-Thorns Starfish in vast numbers were eating out the coral from entire reefs. Intensive research over the next few years revealed that over-fishing in the north disrupted the ecosystem by removing predators that ate starfish larvae. The result was that starfish survived in plague numbers on some reefs. When they bred, ocean currents, moving from north to south, carried their larvae to other reefs, setting off starfish outbreaks there too. Like an atomic explosion, the outbreaks followed one another in a chain reaction that spread southwards down the coastline over a period of several years.

Many other systems also behave like the atomic explosion. A wildfire starts when a piece of fuel catches alight. Suppose this fuel is a tree. Heat and flames from the burning tree set fire to other trees around it. Then they set fire to more trees. Just like the atomic explosion, the fire is a growing network in which the nodes are trees and ignitions create links between them.

The above events are further examples of the domino principle, which we met in Chap. 9. Epidemics spread in the same way. One person catches a cold and infects others. They infect other people, who then go on to infect yet more people in turn. All of these examples, plus many more, are also cases of what is known as *percolation*. In general this is any process that spreads from node to node across a network. There are many other examples: the firing of a laser, water freezing, the spread of introduced pests across the landscape, and the spread of a rumour, to name just a few.

In science, mathematics has been the most common way of revealing similarities. Both compound interest and reproduction, for example, are cases of exponential growth. Throw a stone and its path follows a parabola, which is the same shape you get if you slice through a cone. But mathematics is not just about numbers. It is about abstracting underlying patterns. In the case of the starfish, deep similarities with other systems become obvious if we interpret them as networks. Throughout this book I refer to the networks that underlie human events and relationships. But networks are all around us. And those networks have properties that are responsible for things that happen in many apparently unrelated systems.

Knowing that all of these different events follow an underlying pattern is not mere analogy. It allows us to make precise predictions about them. For instance, most people have heard the term critical mass in relation to an atomic explosion. What this means is that there needs to be a certain amount of nuclear fuel present to ensure that the spreading neutrons do hit other nuclei. The other examples mentioned above all have their version of a critical mass. A fire will not spread if the fuel is too sparse or not dry enough for heat from one burning tree to ignite another. The standard way to contain epidemics is to break the connections, so authorities isolate cases to prevent the disease spreading.

## The Tapestry of Fate

In ancient Greek mythology, the Fates (*Moirai*) were three sister goddesses who spun the threads of people's lives, weaving them into patterns of destiny. The first sister *Clotho*, spun the present onto a spindle; *Lachesis*, measured the thread of the past; and the third sister *Atropos*, determined where to cut the thread in the future. Many other cultures and story tellers have adopted the same theme. In Norse mythology, for instance, the sisters are the *Norns*, who weave good and evil into people's lives. Shakespeare portrayed them as three witches, the Weird sisters, in his play *Macbeth* and Charles Dickens borrowed the idea for the three ghosts who visit miserly old Scrooge in his story *A Christmas Carol*.

Though only myths, the vivid image of events unfolding like threads weaving themselves into the design of a tapestry is a powerful and potentially useful analogy. Finding patterns within the networks that underlie events and relationships can provide useful information. Particular kinds of networks have patterns of connections that make their presence felt in the way systems behave. The spread of an epidemic, for instance, arises not from the nature of the individuals involved, but from the way those individuals interact with one another.

As we saw previously (Chap. 3), any sequence of events is like a pathway winding through a network of possible events and consequences. The choices that face us at any given point in time are like the branches of a tree, radiating outwards from the trunk. The sequence of choices we make forms a pathway through the network of all possibilities.

One problem people have in trying to make sense of life's complexities is that it is hard to understand the networks of possibilities that surround us. This is not always so. In some cases, we can see our choices as visual images. Suppose you travel through a town. Then you can draw your route on a map. The path you took is a visual image of your journey. The twists and turns in the path capture the sequence of choices you made when deciding which streets to follow to get from point A to point B.

The exact nature of the network of possibilities varies from case to case, but we can understand some of the issues by looking at dynamic systems that we can picture easily (Fig. 13.1). When the network of possibilities is highly intertwined, the behaviour of the entire system can be chaotic. What often happens is that the system begins by exhibiting *transient behaviour* in which it deviates rapidly from its previous state. However, it eventually falls into a basin of attraction where it stays until outside forces push it out (Fig. 13.2). In a simple system, the trajectory over time will simply stop at an equilibrium, or else cycle through a sequence of states (like perpetual check in a game of chess). In complex systems, the behaviour is more likely to be chaotic. Within that basin, the way the system changes over time will be unpredictable and it exhibits what is known as *sensitivity to initial conditions*.[3] However, its state will remain somewhere within that basin of attraction.

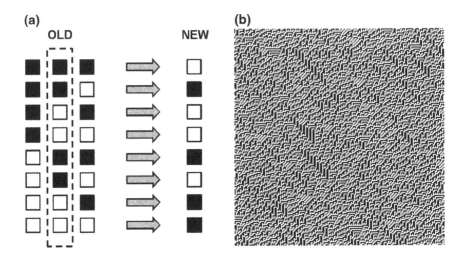

**Fig. 13.1** Patterns of destiny in a machine. This example concerns a row of switches, each of which can be on (*black*) or off (*white*) at any time. **a** The rules governing the ON/OFF sequence for each of the 8 possible combinations of switches and their neighbours. The column labelled "OLD" lists all possible combinations of ON and OFF that can be found for a cell and its neighbours. The *dashed box* indicates the central cell in each combination. **b** Patterns formed as the row of switches turn on and off over time. The pattern of ON and OFF at any time is the state of the machine

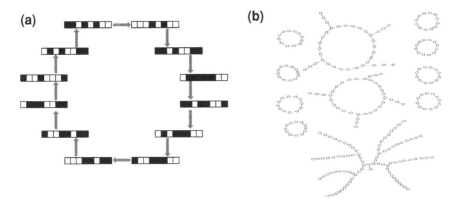

**Fig. 13.2** Patterns of destiny in the machine from the previous figure. Here each state is the pattern of *black* and *white* cells in a row of 8 switches. **a** A set of cell patterns that form a cycle, each pattern is succeeded by the one next to it. **b** The network of all transitions between states formed by the rules in the previous figure. One of the small cycles here is the cycle shown in detail in **a**. Nodes on the outside are "Garden of Eden" states: they cannot be reached if the machine starts in any other state. The sequences of states are transient behaviour and the cycles in the centre are called *basins of attraction* because the machine's state is "attracted" to them. Once its state lies in the basin of attraction it remains cycling within it

Developing the above analogy, we can picture society as a system. Its "state" at any given time is the combination of properties and actions of all the people, its organizations and facilities. If this picture seems too vague or too all-encompassing, then think of something a bit more concrete. Think, for instance, of the flow of money and goods between people and organizations that make up society.[4] In a stable society, all that money, all those goods, will be flowing in a repeating pattern. That repeating pattern is a basin of attraction within which society operates. Major events, such as disasters, wars or the introduction of new technologies, can kick it out of its normal state. For a time it will exhibit transient behaviour: constant change. If the changes produced are great enough, such as events that alter the flow of goods and money, then society will eventually settle into a new basin of attraction. This new basin will involve a new social structure or a new way of doing things.

## The Straw that Broke the Camel's Back

Have you ever noticed that 1 min you've never heard of someone, an actor or singer for instance, and the next minute they seem to appear everywhere and feature in everything? This all-or-nothing nature of fame reflects a curious property that applies to networks of all kinds.

In 1959, two Hungarian mathematicians, Paul Erdős and his colleague Alfréd Rényi were looking at what happens if you start joining up the nodes in a network in haphazard fashion.[5] As we have seen, networks appear in all kinds of situations. They also occur in many different settings, from atomic lattices to coral reefs and almost everything in between.

Erdős and Rényi did this thought experiment. Suppose you start with a set of isolated nodes and add edges to the network in random fashion. You go on adding edges at random until all the nodes are connected. At first, each edge you add will join two nodes together. In other words, they create small clumps of size 2. How does the size of the clumps grow as you add more edges? Intuitively you would expect that the clumps would grow steadily until all the nodes were connected. What really happens is very different. Erdős and Rényi found that a bizarre change, a *connectivity avalanche*, occurs in random networks as they form.

At first the edges join single nodes into pairs. But as pairs of nodes become the dominant arrangement, new edges link pairs to form clusters of four nodes. As the groups of four are twice the size of the pairs, it takes only about half as many new lines before they come to saturate the forming network. Then new lines start forming clumps of 8, 16 and so on, and each doubling of size takes less time than the previous one. Suddenly, it seems, almost all the nodes are linked together into a single *giant component* (Fig. 13.3).

What this appearance of a giant component means is that when you add edges to a set of nodes, they undergo a sudden phase change: one moment the nodes are either isolated or else they are members of small clumps. The next moment they are all connected to a single network (Fig. 13.4). It also means that any random

**Fig. 13.3** The connectivity
avalanche, showing
formation of a giant
component. The numbers
*below* each figure indicate the
number of edges and size of
the largest clump of
connected nodes

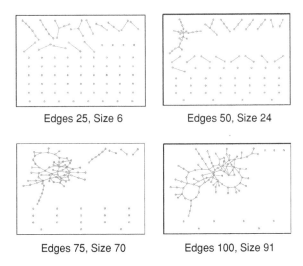

Edges 25, Size 6                    Edges 50, Size 24

Edges 75, Size 70                   Edges 100, Size 91

network that we care to look at will either be fragmented or connected. There is
very little in-between.

The connectivity avalanche is one of the best examples of an event that emerges
out of the connections between individuals, and not from the individuals themselves. It is responsible for many sudden changes that are familiar to everyone. In
physical systems there are many phase changes that occur at a fixed critical point.
Examples include the melting of ice, the formation of crystals from solution, the
nuclear chain reaction; the firing of a laser, the spread of an epidemic, or the
fracturing that follows metal fatigue. In human terms, there is the crowd that
panics, the army that breaks and runs, or the sudden crash in the stock market.
Some networks alternate between connected and disconnected states and self-
organize as a result.[6] This process, *Dual Phase Evolution*, shapes social networks.
We will look at its effects in the following chapters.

## Crisis of Confidence

There are times and places in human affairs where a slight nudge, a single event
can change everything. The assassination of the Austrian archduke in 1914 was
one such event, triggering a series of events that plunged the world into the
catastrophic Great War of 1914–1918.

Throughout history, panic has been a decisive factor in warfare. In a traditional
battle, two armies would approach from different sides of the battlefield, and hack
away at each other. Eventually one side or the other would falter, then crumble and
collapse. The key to understanding this lies in the nature of armies. Before a battle,
soldiers are likely to be gripped by fear. What stops them from fleeing in panic is
group discipline. The need to develop this discipline is the reason why recruits
spend so much time on the drill square.

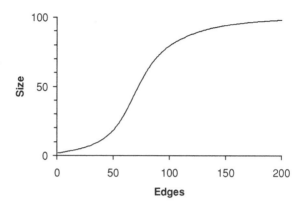

**Fig. 13.4** The connectivity avalanche in a random network of 100 nodes. This shows how the size of the largest cluster increases rapidly as edges are added to connect random pairs of nodes

Being part of a group gives people courage to do things they would never dare to do as individuals. Acting alone, most people would not create a public disturbance, but put them in a gang and they may do all kinds of outrageous things. Put them in an army and they will march across a battlefield thick with flying bullets.

Military courage is a delicate balance between group discipline on the one hand and the desire for self-preservation on the other. Tell an individual to march across a field towards a line of guns all firing at him and he would probably run the other way. But order a regiment to do the same thing and no one dares to be left behind. It is the pressure of the group that makes individuals conform. However, in a dangerous situation, such as battle, there is a delicate balance between discipline and fear.

The critical point comes when the soldiers cease feeling they are part of a fighting unit, and begin to feel they are isolated and vulnerable individuals. When too many soldiers start feeling this way, it only takes one or two to falter and panic quickly spreads.

## The Butterfly Effect

The so-called *Butterfly Effect* is a common feature of complex systems.[3] Thanks to a host of popular accounts in recent years, the butterfly effect is well known. The name derives from the idea that a butterfly flapping its wings in (say) the Amazon could subtly alter air movements, and that the weather can be so sensitive to these small disturbances that it makes the difference between (say) sunshine and a typhoon in China.

The idea that a butterfly flapping its wings could cause a typhoon is far-fetched. For the most part this is true. But as with the nail in the poem at the beginning of Chap. 9, at critical times and places even the tiniest local event can tip the global balance.

As we saw above, the butterfly effect often occurs in human affairs. This idea was explored dramatically in the 1998 film *Sliding Doors*. Two parallel stories trace how the life of Gwyneth Paltrow's character unfolds differently as a result of missing a closing door on a subway carriage. At some time or other, most people find themselves asking "if only ...". My father, a veteran of World War 2, recounted incidents in which he narrowly escaped death because bombs and torpedoes failed to explode close by. It is sobering to realize that without those failures neither my family, nor I would be here.

The butterfly effect arises because the behaviour of complex systems usually *diverges*, rather than *converges*. That is, if you start a complex system running twice, with similar conditions each time, then it doesn't behave exactly the same way; instead its behaviour quickly becomes very different each time. Some examples might help to make it clearer what this means. Let's start with some simple behaviour. If you suspend a ruler by a piece of string and start it swinging, then its motion will soon stop. The ruler will come to rest in a vertical position. The point where the ruler comes to rest is called equilibrium. If you nudge it, or blow on it, then it will quickly return to that position as soon as you stop. This property means that the equilibrium is stable. Now suppose that you manage to balance the ruler on its end. Again, the balanced position is equilibrium, but it is unstable. Any slight nudge will send it tumbling away from its original position.

Complex systems are often far from equilibrium. This means that even when left to themselves, they do not remain static, but rapidly diverge from their starting point. This property makes complex systems sensitive to small changes in their initial conditions.

This property, *sensitivity to initial conditions*, was first noted independently by several researchers. It is one of the hallmarks of chaos. The meteorologist Edward Lorenz, for instance, attempted to simulate weather systems on an early computer. He found that when he restarted a simulation, using results saved previously, the outputs rapidly diverged from the earlier results. After a handful of iterations the model behaved in a completely different way. The difference was explained by small changes that were introduced when numbers in the model were rounded off for printing.[7]

In other words, weather is extremely sensitive to initial conditions. Change the local conditions by just a little bit, and weather patterns in a few days time, and hundreds of kilometres away, could be completely different. This extreme sensitivity leads to the idea of the butterfly effect. Maybe, under certain conditions, the weather is so sensitive that even a butterfly beating it wings can make the difference between sunshine and hail.

Extreme sensitivity of this kind is often impossible to prove, or disprove. However, it is now known that weather events at certain times and places do have a widespread influence over entire regional weather patterns. One of the best known is the El Niño phenomenon. El Niño is an abnormal warming of tropical surface waters in the eastern part of the Pacific Ocean. It is part of a process of reversal (from East to West) in air pressure and ocean warming, known as the Southern Oscillation. In years when El Niño occurs, it triggers abnormal weather events in other parts of world, such as drought in Africa, Australia, and South America, and floods in southern USA. Many other recurring weather patterns of this kind have now been identified.[8]

The butterfly effect is not confined to the weather. It occurs in all kinds of complex systems. If you pour hot water into a bathtub containing cold water, then pretty soon the bath water will be uniformly warm everywhere. That is simple behaviour. On the other hand, if you add a few salt crystals to a saturated salt solution, then the crystals start growing and expanding. That is complex behaviour. Instead of local differences converging, like the bath water, in complex systems they diverge, like the salt solution. If you burn the sausages when cooking, the smoke soon disperses uniformly within the air to lessen the smell (any residual smell is retained by surfaces in the kitchen). But if you accidentally place an apple with a rotten spot in a bowl full of apples, the infected spot does not get diluted. It quickly spreads and within a day or two, the entire bowl may be rotten.

## The Big Break

If there is any profession that knows about one thing leading to another, it is the entertainment business. Performers constantly talk about looking for their "big break". Getting recognized in the first place is a matter of serendipity. You just have to be in the right place at the right time. But you can greatly increase your chances by making serendipity more likely. You can make opportunities.

As any aspiring actor or actress will tell you, the point about getting a big break is that once you get it your career is made. Once you appear in one show, your chances of landing another role go up enormously. One thing leads to another. The down side of this is that if you don't manage to get a lucky break, then you might struggle forever for recognition, no matter how good you are.

A few years ago, I unwittingly experienced for myself how this sort of thing can happen. At the time, I was representing my university in a joint project with several collaborators. During one of our meetings, the representative of one of these collaborating groups asked me to present a short talk at a workshop they were organizing the following week. Now it turned out that a newspaper reporter was attending the workshop. At the end of the session, she asked whether I could write up some of my talk as part of a series of features that her newspaper was running. About two weeks later this article ended up as a half page spread. Phone calls soon followed. In quick succession, I found myself interviewed on national TV and appeared in a whole string of radio programs. Many colleagues have had similar experiences.

Similar effects operate in publishing. Becoming a successful novelist is notoriously difficult. Even to become a published novelist at all is an extremely arduous and competitive business. Many famous novelists have struggled to get their first novels published, even ones that later turned into runaway best sellers. For instance, it took a year of searching to find a publisher for J. K. Rowling's original manuscript of *Harry Potter*, which later grew into a billion dollar industry.[9]

One reason for the above problem is the sheer number of people trying to write novels. However, a more fundamental issue is that publishing is a business, and the first priority of any business is to make money. Publishing work by an unknown author is a lottery. There is a chance that a first time novel may become popular, but more often than not it will be a flop, no matter how well written it is. So publishers prefer safe bets, and the most likely winners are authors who are already successful. Write one best seller, and your other books, good or bad, are likely to sell as well. So the shelves of bookstores tend to be filled with books written by a handful of internationally famous names. Readers flock to buy the latest work by popular authors such as John Grisham, Stephen King, or Wilbur Smith. Publishers tend to sell their names, not their stories.

Extending the principle that famous names sell books, another route to becoming a successful author is to achieve fame (or notoriety) in another way. Almost any kind of fame will do. The writings of sports stars, TV personalities, or politicians are more likely to be autobiographies than novels, but their books sell in huge numbers. Having no ability as a writer is no impediment. Publishers are willing to invest huge sums for any topical story that a ghost writer can make readable.

The movie industry applies the same principle to film-making. Every so often a director with a quirky, original script makes a big screen hit. For the most part, however, films follow sure fire formulae, such as car chases and explosions. Sequels or remakes of popular films are usually safe bets. Big name stars are important draw cards too. The actor Marlon Brando, received $10 million for just 10 min on screen in the 1978 movie *Superman*.[10] The lure of a big name also seems to be playing an increasing part in determining who gets into movies. There are many examples of famous sports stars, models and pop singers moving on to successful acting careers. There are increasing numbers of acting dynasties.

## Avalanches and Cascades

Suppose that a company is preparing its financial plans for the coming year. Typically the directors will consider all the sources of income and expenditure. They might, for example, try to predict annual sales from the market volume and the amount they spend on marketing and advertising. This approach will work fine most of the time. But what if some event occurs that is outside their financial model? What if, for instance, someone introduces an entirely new technology that renders their product obsolete?

In effect, events such as the appearance of a new product create new connections within the network of all possible events. The result is that sets of events, previously isolated from one another, suddenly become connected. Links that were missing from the model suddenly become important. Take another example. In

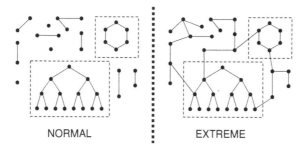

**Fig. 13.5** Cascading contexts. Here the *dots* (*nodes*) represent possible events that can occur in some system (such as in Fig. 7.2). The *lines* (*edges*) connecting the *dots* represent pathways by which one event can set off another event. The *boxes* represent models that treat sets of inter-connected events as closed system, isolated from outside influences. In normal conditions (a "local phase," shown at *left*), the assumption that the systems are closed is valid. When extreme conditions occur (a "global phase," shown at *right*), many new pathways form between events. The models break down because the events inside the *boxes* are now connected to events outside

normal conditions, cars driving along a road conform to the model laid down by the road rules: everyone sticks to their lane; no one stops suddenly and so forth. So no collisions occur. However, events outside the model can change conditions. Suppose something out of the ordinary occurs, such as a tyre blowing or a truck clipping a stone that flies up and breaks a car's windscreen. Then the model of normal driving fails. A whole new set of events become possible, especially accidents (see Chap. 7).

Expressed in terms of networks, a "normal" condition consists of a network of possible events that is disconnected (Fig. 13.5). But unusual events can change that. The system suddenly experiences a phase change: 1 min events are like a closed box, separate from everything on the outside; the next minute there are connections to large numbers of events.

We have seen that some systems change suddenly, from disconnected to con-nected, more and more edges are added between their nodes. Not only does this phase-change occur, it can also have profound effects on the systems concerned. Most of us are not aware of it, but this sort of change happens to most of us almost every day.

Essentially it works like this. We divide our lives, our activities, and our world, into separate compartments. We do this because it makes things simple. By dividing a big company into separate divisions we simplify the jobs that each manager and each worker has to do. We cut out complex chains of tangled problems that can waste days of pain and sweat to sort out. By setting rules and regulations, we make sure that people follow the straight and narrow. By elimi-nating extremes, we ensure that most people, most of the time, need deal only with a smaller, simpler range of matters.

The problem, as we saw earlier, is that all of our efforts to simplify our lives are based on models, which omit many things about the world around us. If the road rules insist that you always drive on one side of the road, they are applying the model that if everyone sticks to one side of the road, then head-on collisions cannot occur. But the world is really more complex than any of our models. The road rules do not take into account extreme conditions, such as brakes or steering failing.

What extreme events do is connect "nodes" that are normally kept well apart. When that happens, they create new, complex pathways that lead to unpredictable and often undesirable outcomes. To see how connecting different contexts can create massive problems, we conclude this chapter by looking at an incident where a series of events and issues, each arising from different contexts, led to disaster.

## Horror at Tenerife

On the afternoon of Sunday, March 27, 1977, a chain of events in the Canary Islands cascaded in dramatic fashion. Seen in its own context, each event was just a minor glitch, but each opened the way for the next and they linked together to create a catastrophe. The result was the worst aviation accident of all time.[11]

At first things were merely inconvenient. At 1:15 p.m., a terrorist bomb planted by a separatist movement exploded in the concourse at Gran Canaria airport. The attack forced the closure of Gran Canaria, the Canary Islands' main airport. As a result, air traffic was diverted to a secondary airport at Los Rodeos. This airport was much smaller that Gran Canaria and ill equipped to handle heavy air traffic. To make things worse, it was a Sunday, so only two controllers were on duty.

As one large airliner after another diverted to land at Los Rodeos, the airport became so crowded that finding spaces for them was like a jigsaw puzzle. Planes were soon forced to park on the taxi way. This meant that planes had to taxi along the runway, which they shared with planes taking offs and landing. The increased danger was compounded by fog, causing poor visibility for both pilots and traffic controllers and the airport had no radar.

The delays caused by diverting from Gran Canaria, and by the congestion at Los Rodeos, put considerable pressure on the pilots of KLM Royal Dutch Airlines Flight 4805. KLM allowed its pilots to fly only a fixed number of hours per day, and the crew was in danger of exceeding that limit. The pilot, Captain van Zanten, was anxious to take off as soon as possible. To save time, he had his plane refuelled during the delay, instead of waiting until it returned to Gran Canaria. So its fuel tanks were full when it taxied down the runway shortly before 5 p.m., in preparation for take off.

While the KLM 747 was taxiing to the end of the runway, the controllers instructed Pan Am flight 1736, another 747, to begin taxiing and to take the third exit on the left. However, after passing the second exit they were unsure of their position and missed the third exit. Neither plane could see the other, nor could the control tower.

At this point a series of communication problems triggered the accident. On reaching the end of the runway, the KLM plane turned and asked for clearance. The tower radioed back with a clearance for the route it was to follow after take-off, but not for the take off itself. When the co-pilot responded with "we're now at take off", the tower replied with a non-standard response of "OK", which the pilot took to mean clearance to take off. The controller almost immediately added "stand by for clearance to take off". But the message was blocked by interference when the Pan Am crew made a simultaneous call to the control tower. This call was to inform the tower that they were still taxiing, but it was also blocked by interference.

When the tower then asked the Pan Am crew to "report when runway clear" they replied "OK, we'll report when we're clear." Hearing this, the KLM flight engineer expressed concern that the Pan Am flight might still be on the runway. However, the captain believed he had clearance to take off and over-ruled him.

As the KLM flight accelerated down the runway, the pilot of the Pan Am saw the approaching lights and attempted to turn. At the same time the KLM captain tried to avoid the collision by powering up to climb faster. However it was too late. At 5:06 p.m., the bottom of the KLM fuselage ripped away the centre of the Pan Am fuselage. The KLM plane then stalled and rammed into the ground about 150 m down the runway producing an enormous fireball.

In all, the crash claimed 583 lives. This included all 234 passengers and 14 crew members on the KLM flight, as well as 326 passengers and 9 crew members from the Pan Am flight. The ensuing enquiries led to claims and counter claims about who was to blame for the accident. Ultimately, however, it is a tragic example of the way a chain of minor problems in different contexts can escalate into disaster.

## End Notes

[1] Saying, known as Poulsen's Law.

[2] Raymond (1986).

[3] Sensitivity to initial conditions (often called the "butterfly effect" in popular writing) refers to systems in which a small change to conditions in the system at the start leads rapidly to different behaviour. For further explanation see Bossomaier and Green (1998).

[4] Remember that this picture is itself a model, so the connections are still not as rich as those in the real society they represent.

[5] Erdős and Rényi (1959).

[6] Green et al. (2006c), Paperin et al. (2011), Abbass et al. (2013).

[7] Lorenz (1963).

[8] NOAA (2013). NOAA's El Niño Page. National Oceanic and Atmospheric Administration. http://www.elnino.noaa.gov/ Downloaded 25 July 2013.

[9]Recounted in J. K. Rowling's biography on her official web site: http://www.jkrowling.com/.

[10]Source: *Trivia for Superman*. Internet Movie Database http://www.imdb.com/title/tt0078346/trivia.

[11]This reconstruction is based on three published accounts: Bruggink (2000), Gero (2003) and Walters (2000).

# On Being Well Connected

# 14

*A most suitable connection everybody must consider it, but I think it might be a very happy one.*[1]

**Abstract**

Links between people form networks of connections. These social connections have always been important in society. Patterns of connections go hand in hand with the nature of peer influence involved. Families, power structures and other hierarchies lead to social networks with tree structures. Traditional social networks are constrained by time and space and form patterns known as Small Worlds. New social media have no such constraints and display patterns known as Scale-Free Networks. New technologies are fragmenting social networks by eliminating the need for interactions between people. A growing social problem is that traditional support networks are disappearing, as highlighted dramatically by instances of people dying alone.

## Degrees of Separation

In Britain during the 1920s, there was a popular song that began: "I've danced with a man who's danced with a girl, who danced with the Prince of Wales." There has always been a great deal of prestige and status to be gained by being well connected. By being close to famous people, some of their fame and some of their influence rub off on us. That's why everyone wants to meet a hero; why fans queue up to get the autograph of celebrities, and why people start name-dropping during conversations.

Another form of snobbery by association is to claim ancestry from someone famous. If you trace your lineage back far enough, you're almost certain to come up with some famous (or infamous) relative in the past. This is because the number

D. G. Green, *Of Ants and Men*, DOI: 10.1007/978-3-642-55230-4_14,
© Springer-Verlag Berlin Heidelberg 2014

of ancestors increases by a factor of 2 for each generation back you go. So you have 2 parents, 4 grandparents, 8 great grand-parents and so on. Go back a thousand years (about 40 generations) and the doubling implies that you have about 1,000,000,000,000 ancestors. This is a greater number than all the people who have ever lived in the entire history of the world. The paradox is explained because many of your distant ancestors appear many times in your family tree. What this says is that if you go back far enough, then you are probably related to everybody whose family roots originated in the same country or region. So if someone proudly claims that they are descended from King William the Conqueror (or some other king), then they are probably correct. But then, most people in England could probably make the same claim.

The first European colonies in Australia were convict settlements. For many years, it was widely considered a source of family shame to be descended from a convict. Most people kept the fact well hidden. However, by the time of the country's bicentenary in 1988, attitudes had changed. In the kind of inverted snobbery that is common *Down Under*, it became fashionable to be able to claim a convict in your family tree.

If you go back far enough in your family tree, you are likely to find that you are related to just about everyone else. Just as a certain percentage of the general population is famous or criminal, so too, your family tree will include a percentage of ancestors who were famous or criminal.

A related phenomenon arises in social networks.[2] How many times have you ever met someone for the first time, only to discover that you have a mutual friend or acquaintance? In 1990, John Guare wrote a play (later turned into a 1993 film) called *Six Degrees of Separation*. The idea that he put forward in the play was that there are no more than six steps separating anyone in the world from anyone else. For example if person A knows person B, and if person B knows person C, and person C knows person D, then A is separated from B by 1 degree, from C by 2 degrees and from D by 3 degrees. Likewise, the lady in the song quoted above was separated from the Prince of Wales by 3 degrees.[3]

Remember that the degree of separation includes everybody, from the richest and most powerful to the poorest and most humble. It includes the good and the bad, the brilliant and the insane. It includes people you admire, and people you hate. It includes people of every race, every creed, and every country.

The six degrees hypothesis is difficult to test. That is why it remains a hypothesis. For example, what does it mean to say you "know" someone? Is a single meeting enough? Does someone you have not seen for 20 years still count? Most of us would have a hard time writing down a complete list of everyone we know. And collecting comprehensive data would be difficult.

However, it is possible to test the idea, at least in part, in specialized cases. One example was inspired by the eccentric Hungarian mathematician Paul Erdős, who was one of the most prolific researchers of his time.[4] In the latter part of his life, he became a sort of mathematical nomad, wandering from university to university in Europe and North America. Turning up without warning at the office of some unsuspecting mathematical colleague, he would state, "My brain is open". He would then proceed to help them solve the most perplexing puzzles. When that was done he would move on to the next institution, leaving his colleague to write up the results of their efforts for publication.

One of the legacies left behind by Erdős was the concept of "Erdős Numbers." After Erdős's death in 1996, it became a mark of distinction to have co-authored a research paper with such a renowned genius. Even to have worked with someone who worked with Erdős became a distinction. Erdős himself was Erdős Number 0. Anyone who had co-authored a research paper jointly with Erdős had an Erdős Number of 1. People who had published with those people had an Erdős Number of 2, and so on. The cult status that a low Erdős Number acquired led some authors to fake stories of collaboration with Erdős so as to give their work greater status. Years after his death, Paul Erdős was still a prolific author![5]

A similar idea, which caught the popular imagination, is the *Kevin Bacon Game*. The actor Kevin Bacon has appeared in many films, alongside hundreds of other performers. But how close are others performers to Kevin Bacon? The idea is to work out a "Bacon Number" for any actor or actress. If he or she has appeared in the same movie, then their Bacon Number is 1. If they have appeared in a movie with someone who appeared with Kevin Bacon, then the number is 2, and so on. Researchers in the Department of Computer Science at the University of Virginia set up an online service, called the "Oracle of Bacon".[6] It automatically searches through a database of nearly half a million performers and works out the Bacon Number for any actor or actress whose name you enter.

Now, you might expect that it would need long strings of names to make a link between any two actors. But no, instead, almost every famous actor or actress has a Bacon number no greater than 2. For instance, Audrey Hepburn's number is 2. According to the Oracle:

1. Audrey Hepburn appeared in the film War and Peace (1956) with Vittorio Gassman
2. Vittorio Gassman appeared in the film Sleepers (1996) with Kevin Bacon.

Even stars of long ago usually have Bacon numbers no greater than 3. Long careers make this possible. Some stars have appeared in films over many decades. For example, according to the Oracle, W. C. Fields has a Bacon number of 3:

1. W. C. Fields appeared in Follow the Boys (1944) with Orson Welles
2. Orson Welles appeared in The Muppet Movie (1979) with Steve Martin
3. Steve Martin appeared in Novocaine (2001) with Kevin Bacon.

It turns out that the average Bacon Number is 2.892. Of all the performers listed by the Oracle, 85.7 % have numbers no greater than 3, and only 1 % have numbers of 5 or greater. The maximum Bacon Number recorded is 10, but no one is saying who that is!

The above figures imply that Kevin Bacon is right at the centre of Hollywood action, and he certainly is close. However it turns out that for 912 other actors the average separation is less than Kevin Bacon's. The actor Sean Connery, who has an average separation of just 2.66, holds one of the lowest averages. This means that Connery numbers are on average slightly smaller than Bacon numbers.

**Fig. 14.1** Creation of a small world. At *left* is a regular network in which each node is connected to four neighbours. The diameter of this network is 50 steps, which is the length of a path from one side of the *ring* to the other. Addition of long-range connections creates shorter paths, so reducing the diameter

Social number games like Erdős numbers and Bacon numbers are examples of the six degrees hypothesis in limited settings. That very few actors have a Bacon number greater than 3 certainly seems consistent with the hypothesis. If 98 actors out of nearly half a million have Bacon numbers greater than 6, it is only because of the limited kinds of association that are allowed in the test.

## A Small World

In Jane Austen's novel *Northanger Abbey*, the heroine and her aunt find themselves isolated at a soirée. They are isolated because it was considered rude to open a conversation with strangers and no acquaintances were present to introduce them. In those days, if a young man wanted to meet a young lady, he could not simply march up and introduce himself. No, first he had to find a mutual friend who could introduce him to the girl he desired to meet. Modern society may not usually be so strict about first meetings, but when strangers meet for the first time they are still more likely to trust each other if introduced by a mutual friend.

The six degrees hypothesis is not simply about parlour games. It is a practical demonstration of properties of networks. It is an application of social research into so-called *small worlds*. The focus of this research is the nature and implications of connections within a network.

The defining character of a small world network is low diameter. The diameter of a network is the maximum number of steps you need to take to get from one node to another (Fig. 14.1). So the six degrees hypothesis really just says that human society is a small world with a diameter of 6.

How does a rule like six degrees of separation arise? Like many features of society, it emerges unplanned as a consequence of people's everyday behaviour. The process involved is an example of Dual Phase Evolution (DPE), which was mentioned in the previous chapter.[7] Whether you realize it or not, our social lives divide into two phases. Most of the time we interact with the people we already know well: our immediate family, our close friends, and the people we work with on a day-to-day basis. In this sphere of our existence, call it a *local phase*, we strengthen some relationships while others fade away through lack of contact.

Every now and then we enter a *global phase*. We go to meetings, we go to conferences, we go to parties and, in the process, we meet people from all over. Some of these meetings create new social connections that continue afterwards. Most of the time, most people spend their days mixing with people they already know. In the morning they go to work, where they mix with colleagues. After work, they meet up with friends. They talk to members of their family. They wave to their neighbours in the street. In short, they interact within their existing network of local connections.

Sometimes, however, something different happens to people. Their work takes them to a convention, or a meeting with clients, or a visit to another branch. In their social lives they go to parties. They travel, they move house, or someone in the family gets married or has a baby. All of these events involve meeting new people. In doing so, people make new, long-range connections in space and time. It is these long-range connections that reduce the separation between individuals to less than 6 steps and turn society into a small world.

Meeting people from far away creates new long-range connections in space. By meeting with older people, the young make long-range connection in time. And it is these long-range social connections that make society a small world.

This business of switching from nearby to far away, from local to long-range, is one way in which we unconsciously form social patterns. We will see more examples of Dual Phase Evolution at work in social settings below.

Making connections between people is only part of the story in forming social networks. The ways in which people break their social connections are also important. When strangers meet for the first time, what leads them to maintain the relationship? Usually it is because they share something in common, such as a personal interest or a business concern. Of the people we meet as long-range connections, some will provide us with lasting relationships. Others will be only brief acquaintances. No one can maintain regular contact with everyone they meet.

For instance, it may be that they form a relationship based on a single issue. Maybe they meet at a convention for stamp collectors, or just happen to stay at the same hotel. If the single issue is central to their character, say political ideals or religious beliefs, then they may form a lasting relationship. On the other hand, if either loses interest in stamp collecting, there may be no other reason to keep in touch. If they do form relationships on the basis of a single issue, then Dual Phase Evolution leads to social patterns in which people cluster around common concerns. This leads to the formation of tightly knit cliques. Within each clique there are many links within, but there are few links between different cliques.

Of course, there are many other ways for social relationships to form. Many friendships form because people share many things in common. These things may be interests like music or stamp collecting, but they may also be attributes, such as living near each other. The point is they share many things in common, not just a single, over-riding attribute. When this happens, a different kind of pattern appears within the social network that forms. Instead of tightly bound cliques, there are long chains with branches and relatively few loops. In a famous study, Bearman

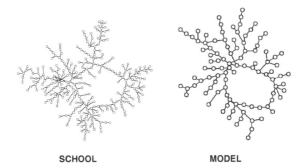

SCHOOL                        MODEL

**Fig. 14.2** A network of social relationships at a high school. At *left* is part of the romantic network found by Bearman and colleagues in their study at Jefferson High School (see text for details). At *right* is a simulated network built to confirm that insights from the Bearman study do explain the observed network patterns. The simulation assumes that Dual Phase Evolution occurs. It assumes that students make random new connections at social events (parties etc.). However, they maintain new relationships only if the pair shares enough interests in common. The two networks share enough similarities—long chains, richly branched in places and no small loops— to provide confidence that the model does explain the social mechanism involved

and colleagues found just this sort of network pattern when they worked out the network of romantic relationships at Jefferson High School (Fig. 14.2).[8]

## How Well Connected is the World Wide Web?

The World Wide Web is a network consisting of pages that are connected by hypertext links. The "diameter" of the Web is 19.[9] This means that you can get from any starting point to any other web page in no more than 19 jumps.

If every page had exactly 3 unique links leading off it, then the total number of sites in the network would be 1,162,261,467. At the time of writing, there are about 800,000,000 websites. However, the distribution of links is far from even. Most sites have many more than 3 links to other sites the richness of connections between Web sites is extremely variable. Some sites are very popular: many sites provide links to them. However, most sites are not widely known or very specialized and have only one or two links leading to them. It turns out that the distribution of links follows a power law. This means that most web sites have only a few links to them from other sites, but a few web sites have large numbers of connections from other sites. In contrast, a simple hierarchy, such as the binary tree that we met in Chap. 4, has three connections per node (one connection up and two down). Just as many actors have appeared in films with popular actors like Kevin Bacon, so too the most richly connected websites form centres about which many other sites form clusters. In turn, sites in those clusters have other sites clustered about them, and so on. From all of these centres, we can measure the separation of other sites.

**Small world**          **Scale-free**          **Binary Tree**

**Fig. 14.3** Three common kinds of networks that arise in social contexts. In a *Small World* (*left*), nodes have local connections, and long distance connections reduce the overall diameter (see Fig. 14.1). Small worlds are typically found in geographically constrained societies. In a *Scale-Free Network* (*centre*), a few nodes are highly connected centres, but most nodes have few connections. Scale Free Networks appear in networks, such as Internet social media that grow by preferential attachment. In a *Tree Network* (*right*), branches of nodes radiate outwards from a single root node (here located at the *centre*). Trees are typically found in designed networks, such as the structure of large organizations

Small world patterns of social connections, which we looked at in the previous section, typify a world in which distance is supreme. It takes considerable effort to make connections from far away (or long ago). The advent of social media forums, such as Twitter and Facebook, has made physical distance almost irrelevant.

When they go online people are not restricted by any limitations of distance. So it does not take any special event for them to meet people from far away. They can chat with someone on the other side of the world as easily as someone in the next room. The social networks that result have different patterns of connections.

What happens in social media is that most people want to be a friend of the most famous, the most popular people around. The result is that some people, the best-known people, acquire hundreds, thousands, or even millions of "friends". This situation leads to a different pattern of social connections, known as a "scale-free network".[10] This means that most individuals have a small number of social connections, but some have extraordinarily large numbers of social connections.

This change in network pattern has consequences. One is that the best-connected people can have extraordinary influence. This has implications for marketing and politics, for example. Convince a few highly influential people to support your product or policy and they may convince hundreds of others.

A network with a distribution of connections like the World Wide Web is called *scale-free* (Fig. 14.3). It turns out that many other systems form scale-free networks. The metabolic system is one.

Scale-free networks have two important features. One is that they are independent of scale—remove any number of nodes and the overall structure still looks much the same. The other is that they are highly robust against random "attacks". If you pick out a few nodes at random and remove them, then the entire network remains connected. If you do the same thing to a tree, then it immediately breaks apart into separate trees. Hierarchies are very brittle. In contrast, a scale-free network can retain full connectivity even when large numbers of nodes are removed. On the other hand, if the attack is selective, if it takes out the most highly connected nodes, then it will soon begin to fragment.

Why are these scale-free networks so common? Why do they form spontaneously in so many different systems? Besides the advantages outlined above, another important reason is the snowball effect. Take the Web as an example again. If a Web site is popular, then people will create links to it. But the greater the number of links that people make to it, the greater its popularity will become. This in turn will encourage yet more people to create links. As we saw in Chap. 3, this is a case of feedback driving the process. It leads to an exponential distribution in the numbers of links to different nodes.

In other words, the presence of a scale-free network suggests that the snowball effect (positive feedback) may be involved during the formation of the network. What this means is that scale-free networks are a form of organization created by complex processes. If you start with a random network and let the system run, then after a time the snowball effect will turn it into a scale-free network.

## Who You Know, Not What You Know

The upstart pretensions of a young woman without family, connections, or fortune. Is this to be endured! But it must not, shall not be. She has no money, no connections, nothing that can tempt him to—she is lost forever.[11]

George is getting worried. He recently finished his studies and is now out looking for a job. His grades were not so good and though he's had interview after interview, he has been knocked back every time.

He recently applied for a job with one of the town's most prestigious companies. He knew that competition would be fierce and thought he had no chance at all. So he is pleasantly surprised when he gets called in for an interview with the boss. But of course, he has been turned down for jobs many times before, so it's understandable that he is very nervous by the time he enters the boss's office.

After the usual polite introductions the boss comes straight to the point.

"OK George," he begins "tell me about your background."

George launches into a well-rehearsed story about his education, his knowledge and experience. Part way through George's spiel, the boss interrupts him.

"So you studied science at school, not economics?"

"Ye-yes." George stammers, worried that this question might mean his chances have sunk already.

"Where was that?"

George tells him the name of his old school.

"Oh well, then you must know Mr. Kepler."

George's nervousness suddenly evaporates. He does know Mr. Kepler, his year ten science teacher. And that means he has something in common with the boss, something important. They are both old boys of the same private school. George feels a warm glow wash over him as he realizes that the job is in the bag. The old boy network has opened the door to a great career in the town's top company.

Connections are everything. From advantageous marriages to plum job opportunities, it is not *what* you know but *who* you know that counts. This was true in Jane Austen's time. It is still true today. If you know the right people then anything is possible. If not, then struggle as you might, success may remain just out of reach forever.

In a recent book, Pamela Laird showed that many of the most famous names in American business owe their success as much to social connections as to personal ability.[12] She also considered other cases, in which people (especially women and Afro-Americans) have had their ambitions thwarted for want of appropriate social connections.

Social connections continue to be an important ingredient of success. In Europe, for instance, nobility is supposed to have lost its power, and yet a recent study[13] showed that in Twentieth century Holland, members of noble families held a much higher proportion of elite positions in society than the rest of the population.

The importance of social connections is not restricted to jobs and marriages. It extends into almost every sphere of activity. Business partnerships are traditionally built on personal networks. Doing business with people you know and trust is usually safer than risking the unknown. And if no one in your circle can do the job, then more often than not you know someone who knows a chap who can. Clubs and business lunches are not just perks of privilege: they also grease for the wheels of industry.

Politicians know better than most the importance of personal connections. The local representative, who knows all the right people, attends all the right functions and knows most of the voters on a personal basis can be virtually impossible to vote out of office.

## Dying Alone

On February 20, 2006 police in suburban Sydney made a grisly discovery. On entering a flat on the 15th floor of a high-rise housing block in the suburb of Waterloo, they found the dead body of a man sitting at his kitchen table.[14] The body had been there for almost eight months and was reduced to a skeleton. The dead man, Peter "Terry" Newman, was a pensioner living alone. No one reported him missing. Police were called by the Housing Department. Fire safety officers carrying out their annual inspection had twice been unable to get any response when they knocked on his door.

The discovery of Mr. Newman's body was not the first case of its kind. Just days earlier, a decomposed body had been found in a housing development in Surry Hills, an inner-city suburb of Sydney. Days later the body of a woman was found at Umina on the New South Wales Central Coast, six months after she had died. And following news reports of Mr. Newman's case, the concerned neighbour of an elderly couple in Neutral Bay called in police, who found them both dead.

In recent years, cities around the world have experienced many cases of people dying alone at home and not being discovered until long afterwards. One of the most bizarre cases was a woman in a London flat whose body lay unnoticed and unmissed for more than 2 years.[15] When found, her TV set was still running and she had Christmas presents under a tree. No one thought it suspicious that there were heaps of letters in her mailbox. She was found eventually, but only because arrears on her rent had become so high that a housing officer and a locksmith forced entry to try to repossess the property.

Sad, extreme incidents like these provide sensational evidence of increasing social isolation. They support claims that social isolation is increasing in modern society. They also demonstrate, in emphatic fashion, that social isolation can wreak damage on individuals.

The problem of social isolation is increasing. More and more people are living alone. In the United States, for instance, census figures show that the percentage of single person households increased from 9 % in 1950 to 28 % in 2011. In most countries of the industrialized world, more than a quarter of all households now consist of a single individual.[16] In many other countries the percentage is even greater: 31 % in Japan and 34 % in Britain, for instance. The country with the greatest percentage of people living alone is Sweden where they make up 47 % of all households.

Many factors have contributed to the above increases in social isolation. One factor is the side effects of new technologies. New technologies have contributed greatly to leisure time in the home. Today most homes in western countries have a host of labour-saving devices: vacuum cleaners, washing machines, dishwashers, and computers, to name just a few.

In like fashion, services have sprung up to eliminate the need for other household chores. Why run around town paying bills when you can pay by phone, or have it automatically charged to your account? Why go to the bank in person when you can do your banking over the Internet? Why run to the post office when you can send email instead. Why waste time at the supermarket when you can order on line and have it home delivered? Why spend hours preparing and cooking meals? Instead you can heat up an instant dinner in the microwave, or grab a meal from your favourite fast food outlet.

So what has happened to all the extra free time? As we noted previously, one important change has been to allow women to escape from the home and get jobs. This brought in more spending money for families. But it also fuelled the rise of fast foods and other conveniences for families all tired out at the end of the working day. We will look at these effects on families in more detail in Chap. 16.

However, one thing that all the inventions and services do not seem to have done is to increase person-to-person interaction within society. Many inventions have served to decrease the need for social interaction.

As we saw in Chap. 1, technology has reduced the link between entertainment and exposure to social contact. Once upon a time, people needed to go out for entertainment. Now they have a range of services and devices that make it unnecessary.

The effect of many new technologies has been to eliminate human contact altogether. In many commercial activities, not even slivers of social contact are left. As we saw in Chap. 12, many innovations are really about reducing costs or creating new markets for the enterprise concerned. Has the influx of "labour-saving" devices into the home contributed to the health and welfare of the population as a whole. Certainly it helped to liberate women from financial dependency and provided them with time to seek opportunities for self-fulfilment. On the other hand, it also served to reduce social contact within neighbourhoods, with all its flow on effects.

Are the combined pressures of technology and commercialization squeezing social contact, and humanity out of the population? Seen in the most pessimistic light, are people being reduced to raw material to be processed, used and discarded by the corporate machine? We will explore these questions further in Chap. 18.

---

## End Notes

[1]From the novel *Persuasion*, by Jane Austen.

[2]A social network is a set of people ("nodes") linked by relationships ("edges").

[3]The idea of six degrees of separation was inspired by research into small world networks. These are networks in which the number of steps separating any pair of objects is relatively small. For actors, for instance, this condition is ensured by players who have been in many movies over many years. Key references include Milgram (1967); Watts and Strogatz (1998).

[4]For accounts of Erdős' life and work see Hoffman (1988) and Schechter (1998).

[5]Oakland University. *TheErdös Number Project* http://www4.oakland.edu/enp/.

[6]*The Oracle of Bacon*, http://oracleofbacon.org/.

[7]Green et al. (2006c), Paperin et al. (2011), Green et al. (2013).

[8]Proving that two networks are similar is difficult. One method is to analyze the frequencies of small characteristic patterns called motifs. In this case, our model shares several frequency patterns with the high school network (Green et al. 2014). These include an absence of small cycles (especially triangles, which are common in clusters), and the existence of chains (paths longer than 2 with no branches).

[9]Albert et al. (1999). Recent report suggest that its diameter may now be greater than 19.

[10]See Barabási and Albert (1999). In a scale-free network the number of friends people follow a power law. The probability of a person having $N$ friends is

$$probability(N) = aN^{-k},$$

where $a$ and $k$ are constants.

These networks are called scale-free because the pattern looks the same no matter what scale you choose to look at it, the pattern is the same.

It is important to note that a network can be both a small world and scale-free. Simply by attracting connections for all over, popular people inevitably reduce the overall diameter of the network.

[11]From *Pride and Prejudice*, by Jane Austen.

[12]Laird (2007).

[13]Dronkers (2003).

[14]Anonymous (2006).

[15]Gillan (2006).

[16]Klinenberg (2012).

# The Herd Instinct

# 15

*No man is an island entire of itself; every man is a piece of the continent, a part of the main.*[1]

**Abstract**

Person-to-person interactions underlie many kinds of group behaviour, from panicking crowds, to clothing fads, to booms and busts in the share market. The growth of the Internet, for example, led to the dot-com bubble at the turn of the Millennium. Communication provides the glue that keeps social groups together. Peer pressure forces people to conform to group morality. New technologies are changing the ways in which people communicate and therefore the way influence spreads. They also make it possible for groups to cooperate in new ways, as highlighted by spontaneous planning during the Paris riots of 2005 and by the electronic collaboration used by groups, such as SETI@home.

## Don't Panic

At midnight on Monday 17 February 2003, the Epitome Night Club on Chicago's South Side was crowded with about 1,500 dancers. Shortly after 2 a.m., a disturbance occurred. The details are not known. Perhaps a man was harassing a girl and she tried to fend him off. Whatever the cause, it resulted in someone letting off a burst of pepper spray into the air. The effect was instant chaos. As people in the immediate vicinity of the spray tried to get away, their panic spread through the crowd. Hundreds of people, screaming and gagging from the spray, all rushed for the doors at once. Within seconds, the doors became blocked and in the massive crush, bodies began piling on top of bodies. Those buried under the stack were crushed. Many died from head injuries or suffocation. By the time the horror night was over, twenty-one people lay dead and dozens more were injured.[2]

D. G. Green, *Of Ants and Men*, DOI: 10.1007/978-3-642-55230-4_15,
© Springer-Verlag Berlin Heidelberg 2014

Panic, spreading through groups of people, is a serious problem in many other situations. As we saw in Chap. 13, panic has always been a danger for armies during battle. Discipline holds the group together in spite of an advancing enemy. Each soldier gains courage from the safety of the group. Where a lone individual would cut and run, group solidarity enables soldiers to overcome their fear. However, if the group shows signs of weakening, such as being forced to retreat, then fear can overcome confidence. Panic spreads quickly and a retreat turns into a rout.

Stampedes also happen in commerce. Share market trends are social phenomena. It often happens that when the price of a stock starts showing an upward trend, buyers jump in and the price rises rapidly. However, buyers want to make a profit, so when the price starts to fall again, people panic and sell. Their panic spreads and the price can crash.

## The Madness of Crowds

Historically, cycles of boom and bust have produced many financial "bubbles". A Typically a bubble goes like this. First, the crowd latches onto some new opportunity to get rich quick. Fuelled by greed, the bubble grows rapidly. Eventually its size exceeds anything that can be supported by financial reality. At this point doubt sets in amongst the more savvy investors. Then some incident pricks the bubble. It may be as tiny as a few nervous investors unloading their stock. Doubt turns to panic and investors start unloading stock in droves. The bubble bursts and prices crash. The unlucky ones, the ones who were too slow to divest their stock wind up out of pocket. If they invested too heavily, they wind up ruined. We saw two historical examples of bubbles—the South Seas bubble and the Mississippi Madness—when we looked at the snowball effect in Chap. 10. Another infamous example was "tulipomania", which occurred in Holland during the Seventeenth Century.

## Tulipomania (1634–1636)

Tulips were introduced into Europe from Constantinople. Their first recorded appearance was at Augsburg in 1559.[3] In subsequent years, tulips grew rapidly in popularity. By the early 1600s, possessing a collection of tulips had become *de rigueur* for the wealthy upper classes in Holland and parts of Germany. As the popularity of tulips increased, so did the prices people were prepared to pay for them.

By 1634, tulips had become an all-consuming obsession in Holland:

> … ordinary industry of the country was neglected, and the population, even to its lowest dregs, embarked in the tulip trade. As the mania increased, prices augmented, until, in the year 1635, many persons were known to invest a fortune of 100,000 florins in the purchase of forty roots.[4]

The demand for tulips continued to grow. In 1636 tulips became a commodity on the stock exchanges of the main Dutch cities. Speculation on the rise and fall of tulip prices resembled gambling. Lured by the prospect of making a fortune, people flocked to the markets. At the height of the mania, in November 1636, almost everyone seemed to be dabbling in tulips. Money poured into Holland from all over Europe, and the costs of basic commodities rose.

The growth of this bubble depended on people being willing and able to pay the rapidly increasing prices demanded for tulips. And they would only do that so long as they could reasonably expect to sell them for a profit. In February 1637, traders found that buyers were no longer willing to pay exorbitant prices for tulips. Realizing that they were not going to make huge profits on the tulips they had already bought, people became worried that they might not be able to sell them at all. The only way to sell was to offer lower prices. But once prices started falling, there was no stopping the trend. The market collapsed. Like Jack selling the family cow for a handful of beans, some families found they had traded all they possessed for a handful of worthless bulbs.

## The DOT-COM Bubble

One of the most influential technologies of our time is the Internet.[5] It grew out of ARPANET, which was established in 1969 as a way to enable military contractors to share information rapidly. The Internet is really a network of networks. The introduction of a suite of internetworking protocols in 1982 allowed many separate networks around the world to be linked together in seamless fashion.[6] Thus was the Internet born.

By 1990, the Internet was already being used world-wide for communication and data exchange. However, you had to be something of a computer programmer to use it. You had to type in complicated commands to get it to do anything. And there were different protocols for everything. For instance, you used Telnet to log into a remote computer and the File Transfer Protocol (FTP) to transfer files.

The hunt was on for a simpler way of exchanging information over the Internet. In 1991, the University of Minnesota introduced a system called Gopher. It provided a system of menus to help students find information on the University's servers. Gopher caught on and became a world-wide phenomenon almost overnight.

Meanwhile, CERN in Switzerland had developed another system, the Hypertext Transfer Protocol (HTTP) for information exchange. Although its hypertext approach was more flexible than Gopher, it was awkward to use in a text-driven interface, so was slow to catch on. A huge breakthrough came early in 1993 when the National Centre for Supercomputer Applications (NCSA) released a screen-driven browser for HTTP. Called Mosaic, this program unleashed the full hy-permedia potential of the World Wide Web.[7] By that time I had been running servers for both Gopher and the Web for more than a year, but dropped Gopher and henceforth ran a Web service only.

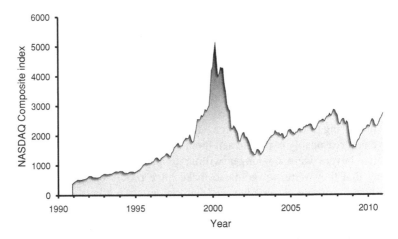

**Fig. 15.1** Fluctuations in the NASDAQ composite index over a 20 year period. The spike in March 2000 reflects the height of the internet bubble

The simplicity of the World-Wide Web opened new possibilities. The development of Web browsers for PCs and the Mac opened the Web up to everyone. By 1994 commercial activity was already underway by companies that saw the Information Superhighway[8] as their pathway to fame or fortune.[9] A notable example was Amazon, which opened in 1994 to sell books over the Web.

Over the next few years, commercial activity on the Web expanded rapidly. Among the most successful enterprises were search engines. On 12 April 1996, shares in Yahoo tripled in price during their first day of trading. Google was incorporated on 4 September 1998. It did not take long for services providers to recognize the potential of popular web sites to gain revenue advertising. As the user base grew, the advantages of moving some aspects of corporate business, such as ordering products, online became very attractive.

The growth of online business is reflected by the rise in the NASDAQ composite index (Fig. 15.1). By the late 1990s, it was obvious to everyone that there were fortunes to be made from internet business. All you had to do was to float a dot-com business and investors would swarm to it like bees around a honey pot. Fuelling the buying frenzy were stories of fortunes being made online. In November 1998, America Online announced plans to buy the Netscape Web browser for $4.21 billion. In February 1999, Forbes Magazine announced that with a fortune of over $10 billion, Jeff Bezos, the founder of Amazon, had become the world's 19th richest man.

By the turn of the Millennium the growth of dot-coms had become an enormous financial bubble. Concern was growing that the value of dot-com stocks far exceeded the true worth of the companies concerned. And yet the bubble kept growing. The NASDAQ composite index doubled within a year. In March 2000, it peaked at 5,132.52. This was the first time it had ever risen above 5,000.

Perhaps it was the significance of this historic high that made investors wary, because at that point the dot-com bubble burst. Over the next few months, several dot-com companies collapsed, alarming investors who were already wary. Twelve months later, the NASDAQ had fallen to half its peak value. It continued to fall for two more years, bottoming out in February 2003 at just a quarter of its peak value. Losses amounted to around two trillion dollars.

When the dot-com bubble burst, the social impact was not confined to a financial crash. The side-effects were felt more widely. One side effect was in training. Lured by the prospect of a booming internet industry, students flocked into tertiary courses in computing, information technology and related fields. Universities were hard pressed to cope with the demand. When the bubble burst, students deserted these career paths. Almost overnight student demand crashed and many institutions had to lay off staff.

## Boom and Bust

The joke goes that an investor is eavesdropping on a financial guru over a noisy line. The guru says: "Looks like it's goodbye Acme". The eavesdropper mistakes this for "Buy Acme" and sets off at a rush to buy Acme shares. Alternatively the guru says: "Acme shares are excellent." The eavesdropper mistakes this for "Acme shares—sell" and sets off at a rush to dump Acme shares.

Stock market booms and crashes are human phenomena. They emerge when crowds of investors panic. Desperate to buy before a share price rises, or to sell before it falls, traders panic lest they be left behind. When the price of a stock starts showing an upward trend, buyers jump in and the price rises rapidly. However, buyers want to make a profit, so when the price starts to fall again, people panic and sell. Their panic spreads and the price can crash. Positive feedback plays a hand in this spread: the bigger the change and the more people buying and selling, the more likely investors are to buy or sell themselves.

The spread of automatic trading has not done away with boom and bust cycles. Instead it has served to magnify their speed and size. As we saw in earlier chapters, traders use programs to do their buying and selling for them. These programs buy and sell according to market trends and patterns specified by the traders. At its very simplest, one such rule might be "if the share price rises by x-percent, then buy".

The whole point of using automated trading is to enable traders to respond quickly to market fluctuations. Responding quickly to rises and falls of just a few points can be the difference between making a fortune and losing one. Humans cannot hope to monitor changes in thousands of share prices that occur in microseconds. Nor can they monitor even a single stock 24 h a day. But computers can.

The problem is that simple rules lend themselves to positive feedback. Somewhere, someone, buys a big block of shares and immediately the price goes up. This rise triggers the automated trading program to buy those shares, sending the price up still further. This rise triggers buying by other automated agents, which forces more price rises, and yet more buying. Pretty soon the share price spirals upwards out of control. The same can happen in reverse if the price falls. Of course, the programs used for automated trading are much more sophisticated than I have made them out to be. Nevertheless, there is still a tendency for rapid spirals to occur. One effect of automated trading is that booms and crashes in share prices tend to be faster, and larger than ever before.

## Calling All Rioters

Modern communications technology serves to accelerate many processes of social interaction. In medieval times, for example, it could take weeks or even months for news to spread across a country. Today it can sometimes spread in a matter of minutes.

On Thursday 27 October 2005, a group of ten teenage boys were playing soccer in Clichy-sous-Bois, a suburb to the east of Paris.[10] The boys were from immigrant families, and mostly of Arabian, Turkish and African origin. Shortly after 5 p.m., police arrived to carry out ID checks. Three of the boys ran. Thinking that the police were pursuing them, these three boys, Bouna Traore, Zyed Benna and Muhttin Altun, tried to hide.

Not realising the danger, the boys climbed over a wall into a substation for the local power grid. There they were electrocuted by a transformer. Bouna and Zyed died instantly and Muhttin was seriously injured. The local area was blacked out.

Within hours, news of the boys' deaths had spread throughout the French Muslim communities of the Paris suburbs. The reaction was like a simmering cauldron of frustration suddenly overflowing. Youths rioted in Clichy-sous-Bois. Police moved in and 27 rioters were arrested. The unrest continued unabated over the weekend. Rioters burned 59 cars, and police made a further 33 arrests.

Just as the unrest in Clichy-sous-Bois started to die down a remarkable thing happened. Immigrant communities elsewhere took to the streets. On Monday 31 October, the rioting spread to Montfermeil, and the next day to Seine-Saint-Denis. By Thursday, the unrest was escalating rapidly. That day there was rioting in four towns and 315 cars were burnt. On Friday riots in Île-de-France, Dijon, Rouen, and Marseille resulted in 596 cars destroyed and 78 arrests. On Saturday, the riots had spread all over France and even into neighbouring countries. Between Saturday, 5 November and Tuesday, 8 November, the riots were at their height. On Monday alone, 1,408 cars were burned and 330 arrests were made.

By November 16, there had been rioting in 274 towns all over France, as well as incidents of varying magnitude in at least seven other European countries. Some 8,973 cars had been destroyed and 2,888 people had been arrested. Remarkably only a single death was reported, but property damage was estimated to be over 200 million euros.

Driving the sudden eruption of violence was widespread fury, fuelled by frustration and resentment. According to BBC reports, French society had negative perceptions both of Islam and migrants: "Islam is seen as the biggest challenge to the country's secular model in the past 100 years". Unable to find employment, poverty kept many families trapped in ghettos. Many Muslim immigrants wanted to integrate into the wider French society, but had instead become alienated and angry.

Modern communication played a crucial part in the rioting. Youths used text messages, email and online blogs to arrange meetings and to alert each other to police activity. Individuals were able to circulate ideas and warnings to their friends instantly. For the most part, the riots were neither planned nor coordinated by an individual leader. Nor was there much coordination between groups in different towns or suburbs. Instead, plans of action emerged as a kind of consensus response to suggestions about times and places.

## Follow the Herd

In 1999, a low budget horror movie, *The Blair Witch Project*, became an overnight sensation.[11] Produced by students, it used a novel documentary approach to story-telling, making a virtue out of its amateur footage. Unlike mainstream movies, it was not advertised in the usual way, but relied heavily on promotion via the Internet. Positive reaction by fans fuelled growing interest in the film. Much of the discussion consisted of speculation about whether it was truth or fiction. Eventually it grossed nearly $250 million worldwide.

Peer influence is well known in business. Fads and fashions are common social phenomena. Word of mouth is a common promotional technique. *Viral advertising* relies on ideas spreading like an epidemic through an existing social network. It is well known that some people are more influential and have more social connections than others. Knowing this, marketers can promote a product by targeting key people within a social network, relying on them to spread the message to everyone else.

As we saw in earlier chapters, a social group is a network. Each person has his or her neighbours, and they have their neighbours in turn. The word "neighbour" here means anyone a person communicates with. This includes family, friends and colleagues; and not just the people who live next door. So if we think of each person as a "node" and these social connections as "edges", then we have a network. The point of this is that by knowing the underlying network, we can use what we know about networks to understand how interactions between individuals lead to behaviour of an entire social group.

Take a very simple case. Suppose that there is some issue on which each person can agree or disagree. Do you vote Republican or Democrat? Do you favour capital punishment? More generally the issue can even boil down to questions as basic as "Do I want to belong to this social group?" or "Will I cooperate with my

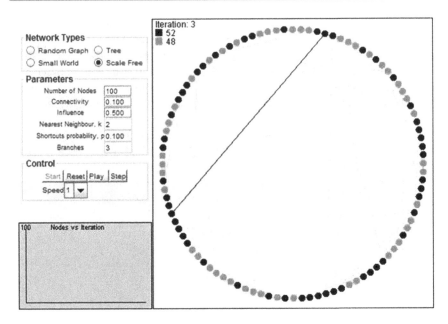

**Fig. 15.2** Setting for simulating person-person interactions in a social network. The *dots* represent people and the *two colours* indicate whether they agree or disagree with a particular idea. Arranging them in a *circle* allows any kind of network to be displayed. The *faint lines* are social connections. The *solid black line* denotes a conversation currently taking place. The control panel at the *left* shows settings used in the simulation. For detail see http://vlab.infotech. monash.edu

neighbours?" For simple binary choices like these, we can view the social group as a switching network in which each person is like a switch: YES or NO, ON or OFF. In this case, interactions between people boil down to a matter of influence about this issue. If person A disagrees with person B, will one of them convince the other to change his/her mind? (Fig. 15.2).

My colleagues and I tested this idea. We reduced the problem down to essentials. People are sophisticated, but all that matters here is whether or not they agree with others. So for the simulation we represented a community as a network of switches, with ON meaning "agree" and OFF meaning "disagree" (shown as different coloured dots in Fig. 15.2). People influence one another through conversations, so we simulated social interaction as a series of conversations, one after another. Again human conversations can be deep and involved, but averaged out over lots of conversations, when two people who disagree hold a conversation, there is a chance that one will change their mind so that their conversation ended with them agreement.

Many would object that the above picture is a gross simplification: people are more sophisticated than this. Of course they are. Social communication can be complicated, very complicated. But that complication can blind us to simple processes that often govern our dealings with one another.

The advantage of the above switching model is that it at once gives us a way to investigate a number of social issues. For instance, as noted in Chap. 5, the British anthropologist Robin Dunbar argued that speech emerged in early human societies as a form of verbal grooming.[12] One advantage of speech is that whereas you can groom only one other animal at a time, you can talk to several others at once. This number seems to be about five. Have you ever noticed that at a social gathering, when the size of a group involved in a conversation gets bigger than five people, it tends to split into two or more separate conversations? One practical reason for this is that it becomes harder to hear above the background noise in a larger group.

As we saw in Chap. 5, Dunbar went on to argue that the greater efficiency of speech allowed humans to form larger social groups than apes. Apes seem to have a natural group size of between 30 and 60 individuals, but the natural group size for humans is about 100–150.

We tested Dunbar's claim by using the model described above to simulate the emergence of consensus in social groups.[13] Our assumption was that a group would fall apart if there was disagreement about fundamental issues, such as whether people agreed to belong to the group. So the question was whether there was an upper limit to the size of a group that could achieve consensus. The results of our studies supported Dunbar's hypothesis by showing that there were indeed upper limits to group size. When the group size exceeded a critical size, consensus was rarely achieved. The critical size limit depended on issues such as the richness of interactions within the group, and the level of influence that individuals exerted on each other.

The above experiment is a simple example of emergence: interactions within the group lead to the appearance of a system-wide property. In this case the property is consensus. However, the emergence of consensus is just one example of a well-known process that occurs in all kinds of systems. It is known as emergent "synchrony", "locking-in", or "enslavement". A striking example of this was shown in a YouTube video of an experiment performed in Ikeguchi laboratory.[14] The experimenters set thirty-two metronomes beating. They were all set to tick at the same rate, but were started randomly so they were completely out of phase with each other. Remarkably, however, within just a few minutes they were all beating precisely in unison. How did this happen? Well the metronomes were sitting on a board which moved slightly in response to the ticking of the metronomes. As more and more metronomes became synchronized this movement became stronger and helped to "enslave" the remaining metronomes in time with the majority.

Synchronization of this kind happens in many different settings. Clockmakers have long known that clocks on a wall will synchronize their ticking (if not the time they show) after a period of a few days. The menstrual cycles of groups of women tend to synchronize after a few months living together. The firing of a laser is based upon enslavement: energy is pumped into the laser and the atoms fire out light of different wavelengths. However as they absorb more and more energy from other atoms, they gradually lock-in on a single wavelength, at which point the laser fires.

I've seen several magicians exploit this phenomenon with their audience. In one case everyone was asked to start clapping, but after just a few seconds everyone was clapping in time. In another case the magician gave the audience a set of words and asked everyone to repeatedly shout out any of the words. He cunningly had some confederates in the audience to shout out a particular word, and after a minute or two the entire audience was shouting out the same word. He then opened an envelope with that word written inside, as if by magic.

As we saw above, much the same thing happens with the way groups of people think. The experiments on social consensus also highlight a danger of decision making by small groups. The danger is that attitudes of member of the group will become locked-in, so they share similar opinions. This is dangerous because the group will then all agree on one approach to any issue. Often the group may be right, but there will be no dissenting view to check them if they are wrong.

## Law and Disorder

Large societies and social groups typically maintain order by prescription: by laws, rules and decrees. Laws are enforced by police and administered by an extensive legal system. But this works only if most people are obeying the law. If everyone stole a car or lawnmower every time they needed one, the police would not be able to keep up. Centrally imposed law works only so long as most people obey the law most of the time. However, there are always incentives to break the law.

Social networks are crucial in maintaining a law abiding society. Peer pressure reinforces views and attitudes that promote cooperation and conformity to community laws and standards. But as we saw above, peer pressure alone is inadequate to maintain order in large social groups. To maintain discipline in large social communities needs a combination of two approaches: prescription imposed from the top and peer pressure acting from below.

Assuming that the above view is correct, any break-down of social networks and personal interactions is a cause for great concern. Without the peer pressure that accompanies social groups, people are more likely to become rebellious and law and order will break down.

My colleagues and I tested this idea using models similar to those described earlier.[15] Suppose we start with a law-abiding community, but there is increasing economic pressure that forces people to behave dishonestly. In the absence of peer pressure, the number of dishonest people (those regularly breaking laws) steadily increases over time. In contrast, in scenarios where peer group pressure played a part, honest people were in the majority. They encouraged those who broke laws to mend their ways, and encouraged others to remain honest. As a result almost everyone was honest most of the time.

However, there were short bursts of dishonesty, where groups of people all become dishonest for a short time. This happened where a local concentration of dishonest people built up and encouraged those around them to become dishonest too. In extreme cases, if the incentive to be dishonest was great enough, then the entire community sometimes flipped and everyone became dishonest! At first we thought it must be a fault in the model, that suddenly everyone could become a criminal. But then we realized that this does happen in reality. On the roads of many cities, for instance, almost everyone tries to drive slightly above the speed limit. Drive at the limit and, in many places, you will get honked and abused by other drivers. In other words, peer pressure works to make everyone a criminal.

We can see this kind of problem in microcosm in the workplace. In an open plan office, everyone works in a single large space. This arrangement encourages interaction. It promotes cooperation and team activity. But at the same time it eliminates privacy, which many people need to concentrate and think. The opposite extreme is a corridor with closed doors leading to individual offices. Such an arrangement allows maximum privacy and freedom to work without interruption on individual tasks. On the other hand it isolates workers from one another, reducing the sense of group identity.

Some years ago I was faced with the problem of group fragmentation when I was managing a team whose members were housed in offices that were scattered across several widely separated buildings. The simple, but highly effective solution was to insist that everyone turn up to morning tea.

## The Eagle and the Ant

In Chap. 2, we saw how ants build an ant colony without planning. Another thing ants do without planning is to search for food. The way ants do this is extremely different from (say) an eagle.

Flying high above the landscape, an eagle has an excellent view of a large area. So all it needs to do is to scan the area beneath it looking for movement, then pounce on its unsuspecting victim.

Ants do not have the luxury of seeing the entire landscape at a glance. Not only are they confined to the ground, they are also very small, so their field of vision (or smell) is very small. Not only that, they do not have brains with which to devise a systematic search strategy. So they wander around the landscape more or less at random until they stumble across food. To avoid getting lost, ants leave a trail of pheromones behind them. They can follow this trail back to their nest. Random wandering of this kind is very inefficient: any single ant is unlikely to find food. The ants counter this problem by sending out hundreds or even thousands of ants scouting for food. In this way they saturate the area around their nest.

As soon as a scout locates food, it returns to the nest, leaving behind a pheromone trail. This helps other ants find their way from the nest to the food and back. Other ants then follow this trail to the food and back. In the process they strengthen the trail, gradually smooth out corners, and make the trail more direct.

The crowd approach to solving problems has been widely used in computing for many years. A good example is SETI@home. There was a world-wide media sensation in 1967 when Cambridge radio astronomers detected signals containing regular pulses coming from outer space. It was suspected that here at last was evidence of intelligent life out there in the universe. However it was soon realized that the pulses were not signals at all, but merely bursts of energy from pulsars. Nevertheless, the excitement about the possibility of intelligent beings sending signals to us helped to encourage a worldwide project called SETI—the Search for Extraterrestrial Intelligence.[16]

The search for extraterrestrial intelligence has always been on the fringes of respectable science. This made it difficult to attract funds to purchase the computer power needed to detect potential signals amidst the wealth of radio data received. To deal with the problem SETI scientists hit on a novel approach. They harnessed the crowd. Literally millions of people around the world are interested in SETI@home, so the scientists provided a free screen saver that people could download and run on their home computers.[17] These screen savers would retrieve data packets from SETI@home, process them and transmit the results back again. By 2001, around 2.6 million computers were engaged in the search.

The approach used by SETI@home is symptomatic of a trend in computing. You can solve a problem much faster by a divide and rule approach: break it down into small chunks and get different computers to work on each chunk. This is like the approach supermarkets use to cope with large numbers of customers. You cannot speed up the rate at which a single checkout can process each customer, but by adding more and more checkouts you can deal with lots of customers at one time. In computing this approach is known as parallel processing and has led to many kinds of technologies, such as computer clusters, grid computing, clouds and multi-agent systems.[18]

In the above examples, including SETI, all the computers involved were doing essentially the same thing: take a part of problem and process it in exactly the same way. But really, this is just a way of applying more brute force. The possibilities are much richer if every processor treats a problem differently, like ants heading off to search in different directions.

Modern communication is helping companies to exploit the *wisdom of the crowd* for innovation. Rather than hire permanent staff and set them to solve problems, a company presents each problem to a crowd and picks out the best solution. Examples of this approach include looking for new designs and producing new apps.

When people look for solutions to a problem, the search is a bit like finding an item of food somewhere in a landscape. Just like a landscape, you can head off in different directions, applying many different approaches. The trouble is that individuals are limited by their knowledge, experience and biases.[19] So they head off in preferred directions, and may miss searching entire regions.

The limitations of individual knowledge and bias pose dangers when decision-making is left up to a single individual, or to a group of people who all think alike. Even worse is a leader surrounded by toadies and yes-men, who just reinforce the leader's biases. Diverse groups of people can do much better. There is wisdom in crowds.[20] Like ants exploring a landscape, a crowd of people with diverse views and experiences will explore any issue from many different perspectives. Like ants heading out from a colony, they will spread out in every direction, testing a branching network of possibilities. In this way, they can find good solutions faster and more reliably than any one individual alone.

Recent research suggests that this "wisdom of crowds" played an important role in the evolution of modern humans and the beginnings of technology.[21] Around 40,000 years ago, there was a bursting forth of creative energy amongst *Homo sapiens* in Europe. Suddenly there were beautiful cave paintings, new stone tools, new ornaments and new implements. The prevailing theory to account for this change was that humans suddenly evolved to have greater cognitive ability. However, accumulating evidence now implies that what actually happened was something quite different. Human beings did not suddenly become smarter. Brilliant individuals had been coming up with new ideas for hundreds of thousands of years. Instead, what happened was like an atomic explosion. Human populations reached a critical density that enabled new ideas to spread...

> Surprisingly early examples of technological and artistic inventiveness indicate that human creativity simmered for hundreds of thousands of years before reaching a boil around 90,000 to 60,000 years ago in Africa and 40,000 years ago in Europe. Social factors, including an increase in population size, seem likely to have amplified our ancestors' powers of innovation, both by improving the odds that someone in the group would come up with a breakthrough technology and by fostering connectedness between groups that allowed them to exchange ideas.[22]

---

## End Notes

[1] Quotation from *Meditation XVII*; John Donne (1572–1631).

[2] Anonymous (2003). 21 dead in Chicago night club stampede. *The Sydney Morning Herald*, February 18, 2003.

[3] This account is based on Mackay (1841).

[4] Mackay (1841), p. 94.

[5] For a more detailed history of the Internet see ISOC (2012).

[6] The most important elements are the Internet Protocol (IP), which specifies how to manage connections between local networks and the Transmission Control Protocol (TCP), which manages communications between hosts.

[7] The term "hypermedia" refers to the combination of hypertext and multimedia. The key to the Hypertext Transfer Protocol is the ability to retrieve information simply by clicking on text (or images). Multimedia means combining many types of media (e.g. text, images, sound, animation) in the one interface.

[8] This catch-phrase was popularized by Al Gore, who was then U.S. vice-president.

[9] I was myself head-hunted by the Vice-Chancellor of a small university who saw the potential of the Web in education.

[10] Wikipedia (2005). 2005 civil unrest in France. http://en.wikipedia.org/wiki/2005_civil_unrest_in_France.

[11]Source: International Movie Database, www.imdb.com. June 2013.

[12]Dunbar (1996).

[13]This research was described in several articles, including: Stocker et al. (2001), Bransden and Green (2005) and Green et al. (2006b, c). For an online demonstration see http://vlab.infotech.monash.edu.au/.

[14]Ikeguchi Laboratory (2012). *Synchronization of thirty-two metronomes.* http://www.youtube.com/watch?v=kqFc4wriBvE.

[15]Green et al. (2006b).

[16]http://www.seti.org/

[17]http://setiathome.berkeley.edu/

[18]See Green (2004), especially the chapter "Talk to my Agent."

[19]See Chap. 4, The glass half empty.

[20]Surowiecki (2004).

[21] Pringle (2013).

[22]Ibid.

# The Subatomic Family

# 16

*It is appallingly obvious that our technology exceeds our humanity.*[1]

**Abstract**

The introduction of new technologies has made enormous changes to family life. During the second half of the Twentieth century, the introduction of labour-saving devices into the family home led to huge social changes. By creating more free time, these devices set off a cascade of flow-on side effects, including increases in married women in the workforce, increased affluence, growth of the fast food industry, and increases in the numbers of child day-care centres, in divorce rates and in the incidence of single parent families. Decreases in social involvement in local neighbourhoods were offset by increased social contact through the workplace.

## One Big Happy Family

In *The Jetsons*, a popular TV series of the early 1960s, the cartoon studio Hanna-Barbera portrayed an American family of the future. The cartoon characters were surrounded by robots, flying cars and other space-age gadgets. Like most visions of the future, the focus was on the material trappings alone. In social terms, the world of the Jetsons was essentially Middle America of the 1950s. What the series failed to anticipate were the many ways in which technology was already transforming society.

The Jetsons were portrayed as a typical 1950s family. Mr. Jetson headed off to work each day; but he used a flying car. Mrs. Jetson was a housewife, but she had robots to do all the housework. People regularly visited other planets at weekends, but still attended local bridge clubs and lodge meetings every week. Even as the

D. G. Green, *Of Ants and Men*, DOI: 10.1007/978-3-642-55230-4_16,
© Springer-Verlag Berlin Heidelberg 2014

series was being aired, the typical American family was changing rapidly. Attendance at lodge meetings, bridge clubs and bowling tournaments was already in decline.

In his controversial book *Bowling Alone*,[2] Robert Putnam bemoaned the decline of participation in community activities across America. During the 1950s, a typical American family would participate in a host of community social activities, such as attending church and meetings of the local parent-teacher association. One or both parents would also take part in regular social activities, such as bowling or bridge club meetings. By the 1990s, however, community involvement in local social activities had waned. Social clubs often struggled and church attendance had fallen dramatically.

Putnam's critics argue that social participation did not die, but merely transformed. One argument is that although people may not talk to their neighbours as much as they used to, a greater part of their social life centres around the workplace. Women may no longer attend the local bridge club, but they do network with colleagues at happy hour after work. On the other hand the increasing focus on the workplace raises concerns about changing attitudes that may themselves contribute to further social changes.[3] Likewise, attendance at local church services may have declined, but tele-evangelists and religious mass movements are thriving. Maybe people don't chew the fat so often with their neighbours, but they do visit chat rooms and exchange stories with like-minded friends online. In other words, technology has transformed the nature of people's social interactions.

A paradoxical effect of better communications has been social fragmentation and isolation.[4] Social contact is a fundamental human need. We are social animals. Technological innovation has contributed relentlessly to isolate people from one another. As the increasing frequency of people dying alone reveals, all is not well in modern society (see Chap. 14). Today there are many new ways in which people are able to lead rich and fulfilling social lives. And yet the social fabric that connected people at the local, geographical level has largely disappeared. And the trend towards social fragmentation just keeps on increasing.

Many technological changes were introduced in the name of efficiency and time saving for the consumer. It is more convenient if you can do your banking online from home. But the unforeseen side effect is to chip away at our exposure to social contact. Perhaps the most serious consequence is that convenient technologies are fragmenting the social networks that reinforce important moral, ethical and social values.

## The Harmless Refrigerator

Social changes often emerge unseen as people adapt their habits to new conditions, new surroundings and new technologies. The introduction of the refrigerator, and other labour-saving devices, provide an excellent example of this.

Until the mid Twentieth century, going to the local shops was a daily ritual in most households. Early in the century, the majority of married women in the western world spent their days as housewives. While their husbands went off to work each day, they stayed at home, looked after the housework and cared for young children. One of their daily tasks was to prepare meals. Food did not keep, so they spent an hour, perhaps more, visiting local shops to buy fresh meat and other staples.

When I was a child, my mother used to spend a significant part of every day visiting the local shops to buy the goods our family needed. For the most part this meant buying perishable food. Some perishables, especially milk, were delivered from a horse drawn cart that passed down our street shortly before dawn each morning. For the rest, my mother had to go shopping. She would go to the butcher for meat, to the grocer for jam and butter, to the baker for bread and to the greengrocer for fruit and vegetables.

In the course of doing their shopping, women would talk to the shopkeepers, as well as many of their neighbours. In this way, an enormous amount of local news and social views were disseminated throughout the community.

The invention of the refrigerator changed things almost overnight.[5] If food could be stored for long periods, there was no need to go shopping every day. It also meant that retailers could stock perishable food. This ability played a role in the rise of supermarkets, which provide virtually every kind of food in a single store.

By the early 1960s, my mother no longer went shopping every day: having acquired a refrigerator, it was no longer necessary. Instead we did all the shopping in a single Saturday morning. And instead of doing the rounds of local stores, we would drive to the nearest supermarket and return home with a car full of groceries.

This new shopping regime was far more convenient. It freed up the time that my mother formerly spent on shopping. The purchase of other labour-saving devices—a washing machine and vacuum cleaner—freed up even more of her time. She took advantage of this free time to find a job. This brought in a second income to our home and made it possible to afford a second car, a new TV and other conveniences.

Census statistics confirm that my mother's story was typical.[6] The decades following World War II saw a host of labour-saving devices become widely available in many developed countries.[7] These included the washing machine, refrigerator, vacuum cleaner, dish-washer, and microwave oven. Whilst the introduction of labour-saving devices may not have been the sole factor that led to more women joining the workforce, there certainly is a strong correlation.[8] In 1933, the percentage of married women aged 35–44 in the workforce stood at a mere 5.3 %. By 1961 the figure had crept up to 21.3 %, but then it increased rapidly. It reached 71.3 % by 1991 and has remained above 70 % ever since (Fig. 16.1; see End note 6).

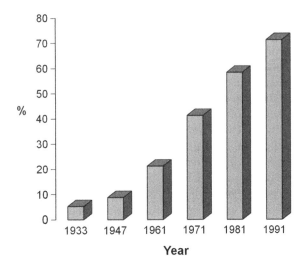

**Fig. 16.1** Increase in the percentage of married women aged 25–34 in the Australian workforce during the 20th century (see End note 6)

Like many new technologies, labour-saving devices were introduced in the name of time-saving and efficiency for the consumer. However, the time and effort they saved for individuals and for families set off cascades of unintended social side-effects. Many of these side effects were not beneficial to the health and happiness of the population as a whole.

New devices such as the refrigerator also had wider social and commercial implications. In the midst of buying food each day, my mother also came into contact with shopkeepers and neighbours. In other words, our little shopping trips were also social outings. She exchanged news and ideas and generally acted as an integral member of the local community. We knew and socialized with our neighbours.

The introduction of freezers and supermarkets eliminated all this. Not only did people not need to visit local shops any more, they could not. One-stop super-markets drove many small stores out of business. Many local stores disappeared. The grocery store became a thing of the past. We no longer knew our neighbours. We no longer had a clear sense of community issues. And we lost touch with many of our friends.

Washing dishes used to be a family event in many homes, but the dishwashing machine made it unnecessary. So what has happened to all the extra free time? Instead of spending it socializing, people collapse in front of the TV for hours each evening. On the other hand, many people find themselves working longer and harder just to stay still. The social effect of these changes is that people do not get out and talk to their friends, family and neighbours nearly as much as they used to do. The change has been noticeable, even within my own lifetime. Besides the increasing social isolation, it also has a detrimental effect on community health. My own country, Australia, used to pride itself as a nation of outdoor, sporting people. Nowadays, surveys indicate that instead of participating in sport, they are more likely to be watching sport.

**Fig. 16.2** Rise in the
number of day care centres in
Australia during the late
Twentieth century (see End
note 6)

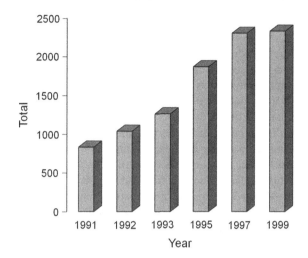

In the book *Future Shock*,[9] Alvin Toffler described the disintegration of human relationships during the late Twentieth century.[10] Human interactions are reduced to brief contacts—talking to the bank teller, buying the groceries and so on. Sometimes, even marriage may be regarded as a temporary convenience. Instead of making lasting friendships, we have temporary slivers of relationships. One effect of automation has been to eliminate even these slivers of human contact.

For many families, labour-saving devices had a surprising and paradoxical effect. What labour-saving devices did was to reduce the time required for house work. More free time made it feasible for both partners to find paid jobs. Perversely, this in turn meant that instead of having more free time, families ended up with much less free time. This shortage of time, combined with extra income, changed people's living habits. Having little time for meal preparation, people looked for alternative ways of obtaining meals. This contributed to the growth of fast food industries during the last few decades of the Twentieth century. Also, with both parents in the workforce, taking care of young children became a major issue for families. Solving this problem led to a rapid rise in the numbers of day care centres in the late Twentieth century (Fig. 16.2).

Admittedly the above interpretation of social changes is somewhat simplistic. Social trends are usually complex and involve many interweaving processes. Certainly many influences played a part in the social changes I have described here. But it would surely be hard to make a convincing argument for the opposite view: that the influx of labour-saving devices had no effect on family life.

In the account given here, I have focussed on Australian trends, but similar trends have occurred in many other countries. United Nations data on the uptake of new technologies[11] imply that similar trends have taken place in most western countries. They are simply on different time scales. At the time of writing, many developing countries are currently undergoing similar patterns of change.

## The Way We Are

The introduction of new technologies during the Twentieth century set off cascades of social side effects that contributed to huge changes in the way modern families live. It is possible that a person transported forward in time from the 1950s would find today's world stranger than the supposedly futuristic one inhabited by the Jetsons.

Today, in most Western countries, for instance, both men and women want to establish a career before marrying. As increasing numbers of women entered the workforce, the problem arose of how to care for young children. Solving this problem led to several changes in people's behaviour. One result is that on average, young couples delay getting married until they are older than in the past.

The introduction of reliable methods of contraception has made it possible to delay pregnancy indefinitely. One effect has been to fuel changes in community attitude which in many Western societies has all but removed the former taboos about sex before marriage and *de facto* relationships. Contraception also means that couples can choose when (and if) they want to have children. Many young people want to save enough money before marrying, and even more so before raising children. So they put off starting a family until they are financially secure. An indicator of this behaviour is an increase in the average age at which women gave birth to their first child (see End note 6).

A further issue is that even after starting a family, both parents often want to continue their careers. Many parents find they need two incomes just to meet the increasing costs of living. So they have to find some way to care for young children while they are at work. In the age of the nuclear family, this means they need to find child care outside the family. As we saw above, the result was that the number of commercial child care services tripled during the 1990s.

Information technology is rapidly changing our mating habits. Recent estimates are that in the U.S., some 16 million people, or about one in four single people, now use Internet dating sites to find partners.

> The previous venues for finding mates – religious institutions, mixers, matchmakers – are being replaced by where the new generation is to be found – at the office, and on the Internet … in the last few years, online dating has switched, becoming something of a destination, not of last but of first resort.[12]

There are also many trends in the way we work. As the number of women in the workforce increased, they make up a growing percentage of the total. In the U.S. women accounted for just 32 % of all workers in 1960, but this figure had risen to 46 % by 2007.[13]

The spread of automobiles during the Twentieth century led to rapid growth of sprawling suburbs around major cities. In my home town, tens of thousands of workers jump in their cars each morning and spend the next hour and a half crawling to their offices in the city. Each evening they repeat the process in reverse. By the year 2009, Americans were spending an average of 25 min travel time to get to work.[14] Over 37 million (33.5 %) were taking over half an hour to reach work and some 3.2 million were spending more than 90 min per day commuting. And all of those commuters had an equally long journey home each day. Attempting to escape the cost of city living, some escape to the countryside, but commute long distances to work into the city several times a week. Sometimes couples work in different cities and commute to be together at weekends.

As cities grow larger and more crowded, simply commuting to and from a distant office absorbs a greater percentage of the day, not to mention the cost and stress involved. If the automobile spreads people out, then the computer is making it unnecessary for them to commute at all. Increasing numbers of people are able to work from home as tele-workers. They use the Internet to pass information back and forth and use video conferencing for meetings. This also makes people more productive. In the United States between 1980 and 2000, the number of workers staying at home almost doubled, from 2.2 million in 1980 to 4.2 million in 2000.[15]

Certainly, household devices such as the refrigerator, vacuum cleaner, washing machine, dishwasher and microwave have saved people hours of domestic labour each day. But they also came at a cost. Dual-income families have greater spending capacity and greater expectations about standards of living. Whole generations of young children are now bundled off to child care centres. Divorce rates, numbers of single parent families and personal debt all soared.

## The Subatomic Family

The family, perhaps humanity's oldest and most important social network, has a tradition of mutual assistance that stretches way back into prehistoric times. Caring for infants, for example, is virtually a full-time task, but the family unit made it possible for one person, usually the mother, to specialize on this task while other family members supported her. However, child care is just one of many ways in which members of an extended family help each other. Cooking, shopping and cleaning are all time-consuming tasks that a family network shares.

Earlier, we saw how an influx of labour-saving technologies into the home has given people more free time. Women, for instance, no longer needed to be tied to the home all day. Instead of being forced by necessity to remain housewives, they could go out and earn money for themselves.

In turn, all of the above trends have contributed to other social trends. In families where both parents have full time jobs, everyone is exhausted at the end of a hard day at the office, so social activity becomes less attractive. It is far easier to grab a take away meal and collapse in front of the TV. Social bonds within the community grew weaker.

Another important side-effect was financial independence. If members of a family have their own income, then they are no longer financially dependent on each other. If a wife has free time to take on a job where she earns money, then she is no longer financially dependent on her husband. Freedom from financial con-straints meant that an unhappy couples did not need to stay together if they did not want to. This change contributed to increases in the incidence of family break-downs and separations.

The spread of labour-saving devices has therefore accompanied, and no doubt contributed to, increases in the frequency of divorce and single-parent families. The U.S.A. experienced a similar change: the percentage of single parent homes increased from 18.5 % in 1970 to 27.7 % in 1999.[16] In other words, more than a quarter of all children in the U.S. are living with just one parent. These increases parallel rising divorce rates: in 1970, just 5.7 % of first-time marriages ended in divorce, but by 1998 the rate had risen to 18.5 %. Another phenomenon was that increasing numbers of single women decided to became "mothers by choice". This includes both adolescent mothers and increasing numbers of older, more affluent (and predominantly white) women, who have elected to become single parents either through out-of-marriage births or by adoption. Similar trends are evident in other countries. For instance, over a 20 years period, the proportion of Australian children living with a single parent doubled, from 12 % in 1986 to 24 % in 2006.[6]

There are ethnic differences in the prevalence of single-parent families. In 1999 the rate of single-parent families in USA was 56 % among black families, 32 % among Hispanic families and 20 % among white families.[16] Several factors contributed to the high percentage of single-parent families among blacks, especially high rates of divorce and young, unmarried girls becoming pregnant and having children.

Trends in the proportion of extended families living together vary greatly from country to country. In the Czech Republic, for example, a housing shortage resulted in some 20 % of families cohabiting with relatives. In most western countries there has been a decline. In the U.K., for instance, the proportion fell from 3 % in 1961 to 1 % in 2004. In Australia, the number of families in which a grandparent was the main carer for a child fell by 39 % in just four years between 2003 and 2007.[17] In South America the percentage has been declining in some countries (e.g. Paraguay, Chile, and Uruguay) but increasing in others (e.g. Brazil, Columbia).[18] In all cases, the figures concerning extended families are complicated by various economic issues, such as the cost of housing.

Although the extended family has declined in most western countries, social behaviour and economic expediency go hand in hand. If the cost of housing and the cost of living go up, then more and more people rely on extended families to make ends meet. One trend has been for young people to marry later and to live with their parents much longer. Even after they marry, many young couples, unable to afford a house of their own, find it convenient to live with parents while they work to earn enough capital to get settled.

# End Notes

[1] Attributed to Albert Einstein.

[2] Putnam (2001).

[3] de Graaf et al. (2001).

[4] See the section on Dying Alone in Chap. 14.

[5] Many labour-saving devices, including the refrigerator were actually invented earlier, but became practical as household items only after decades of research and development.

[6] ABS (2009).

[7] Many of the devices listed here were actually invented decades earlier, but few became available for home use until the post-war period.

[8] With complex changes such as this, it is impossible to untangle all the interacting processes. Another factor was changing social attitudes, especially the rise of feminist movement for women's rights.

[9] Toffler (1970).

[10] Ibid.

[11] UNSD (2009). Industrial Commodity Statistics Database, United Nations Statistics Division, http://data.un.org/.

[12] As mentioned in Chap. 12, Penn (2007) identifies these and numerous other "microtrends" in the way people live.

[13] Source: US Bureau of Labor Statistics (2007).

[14] McKenzie and Rapino (2011).

[15] Source: US Census (2000).

[16] All U.S. figures presented here were obtained from the U.S. Bureau of the Census. http://www.census.gov/.

[17] According to ABS figures, in 2006–2007, there were 14,000 Australian families in which the grandparents were guardians or main carers of resident children (0–17 years) who lived with them. This is a decrease from 23,000 such families in 2003.

[18] http://www.pobronson.com/factbook/index.html

*The more the data banks record about each one of us, the less we exist.*[1]

**Abstract**

Advances in computing and communications have created an information explosion. Information is now an important social and commercial commodity, as exemplified by expressions such as big data and data mining. Instant communications have created a global society in which Western culture is spreading worldwide, placing minority cultures at risk. Mass media have great influence on public opinion; by influencing the choice of what information we see, new media raise the prospect of even greater influence over people's thoughts and opinions. Technology now enables monitoring of individuals to a degree approaching Orwell's 1984, raising concerns over privacy and personal freedom.

## The Information Explosion

One very special treat when I was a child was the day my parents took me to see CSIRAC, an early computer.[2] Built in 1949, it was the only computer in my home town, Melbourne. Occupying a large room, the machine consisted of rack after metal rack, each packed with valves and wires. Most impressive were the panels covered with flashing lights and dials.

And yet, for all the engineering brilliance represented by CSIRAC's vast array of electronics, its computing power pales in comparison with machines today. Its central processing unit ran at 500 Hz; a typical laptop today runs at 3.5 GHz, about 7 million times faster. CSIRAC's memory held about 2,048 bytes (2 KB). At the

D. G. Green, *Of Ants and Men*, DOI: 10.1007/978-3-642-55230-4_17,
© Springer-Verlag Berlin Heidelberg 2014

time I write this, a typical disk drive holds at least 200 GB, or about 100 million times as much data.

The volume of data that computers can store and process has increased just as rapidly. In a project I worked on in 1970 we had to submit data to a large mainframe computer in batches, because the machine (one of the fastest around) could not store more than 1,000 records (about 80 KB) in memory at one time. Forty years later, my laptop computer could easily store 25 million such records on its central memory. In the mid 1980s, my first home computer had disk storage for 20 MB of data. The computer I am now using has more than 10,000 times as much storage capacity.

The very first personal computers (PCs) had no pre-packaged software. Every task required coded commands, which had to be typed laboriously by hand. The developers assumed that users were computing experts. People had to adapt to the requirements of the machine, and not the other way around.

The year 1984 was a turning point. Apple Computers released the Macintosh. Here was a machine designed as if people mattered. You didn't need to be an expert. You didn't need to know a thing about computers. Once you learned how to point and click you could be up and running in a matter of minutes. And people did, in their millions. People who had never touched a computer before became instant converts. It was like a religious conversion. A Doubting Thomas one day became a fanatical apostle the next. Other suppliers soon developed similar interfaces of their own. User-friendly machines changed the PC from a toy into a universal office tool.

One of the many side effects was to make desktop publishing a reality. Previous electronic publishing systems were little more than text editors, with a bit of formatting thrown in. Until you printed a document, you did not really know what result you would get. With a user-friendly computer, you could see what you would get as you typed a page on screen. Errors could be corrected before printing. Publishing software allowed users to design entire newspaper layouts on screen.

The increase in computing power is not the only change that has taken place in the last 50 years. Computers have also evolved into a bewildering variety of species, each adapted to its specialized environment. A few years ago I was asked to speak to a hall full of science students who were just beginning their studies at university. I began by asking who had a computer with them. Only four students put up their hands: the ones who were carrying laptops. Next I asked who had a digital watch, a mobile phone or an MP3 player. Everyone in the hall raised their hands. They all had computers.

The point of my question was that computing has become all pervasive. There is a story (probably apocryphal) that in 1943, Tom Watson, then CEO of IBM, was asked how many computers the world needed. "There is a need" he supposedly replied, "for perhaps five computers in the world." Real or not, the story highlights a massive change in the perception of computers during the late Twentieth century. In the 1950s, only very large organizations could afford to own computers, let alone house and manage them. As computers became smaller, cheaper and more powerful, it became practical to make them available for individuals. By the mid 1980s, the advent of personal computers led to a rapid expansion in the number of users, and in the range of functions they could perform to make jobs easier. Well before the turn of the Millennium, they had become standard equipment, not just in every business, but on every desk.

Computers made their greatest impact on society when people started linking them together. A local area network (LAN), for instance, allows data to be shared by all the computers in an office, so increasing the ability to share and combine data. Communications protocols allow institutions such as banks and airlines to transfer data worldwide, making it possible to withdraw cash at teller machines wherever you happen to be, or to book flights and hotels on the other side of the world.

In the 1970s, the US military began experimenting with DARPANET, which allowed its contractors to share data between their machines. This system formed the basis for the Internet, which today links together almost every computer in the world. The World Wide Web is based on a protocol that simplifies access to data across the Internet.[3] In the early 1990s, shortly after the Web was founded, all the web sites in the world could be listed on a single page. By 2012, there were well over 600 million separate web sites.[4] Today, every respectable organization has a web site to inform prospective clients about its products and services.

The rise of the Internet was part of a phase change in the way people communicate. As we saw earlier (Fig. 12.2), the predominant means of personal communication in 1995 was by fixed phone lines. By 2005, the combination of mobile phones and the Internet had replaced fixed phone lines as the way most people communicate with each other. In just 10 years, society has been transformed.

This incredible rise of computers and communication technology has affected virtually every aspect of modern society. New breeds of computers find roles almost everywhere in our lives, even controls in modern cars, ovens and washing machines. Perhaps the greatest impact of computers has come from changes they have made to the ways we gather, store and distribute data and information. The side effects have had far reaching consequences on society. As discussed in an earlier chapter, the most fundamental change is that we have moved from a world in which data were rare and expensive to one in which they are abundant and cheap.

## Society's Memory Banks

Modern society generates vast quantities of knowledge and information. It all needs to be retained and shared. Fortunately, computers provide the means to capture, store and distribute most of this knowledge worldwide. In 2013, the volume of data created and shared via the Internet was estimated to be around 4 ZB and was tipped to double within 2 years. A zettabyte (ZB) is about $10^{21}$ bytes, or a about a billion terabytes.[5]

We can see the impact of "Big Data" in the way science is done. A good example is the growth of biological information resources. Genbank, the main US repository for biological data, was established in 1982 with entries for 606 DNA sequences.[6] Twenty-five years later, in February 2007, the number of entries had grown to over 10 million sequences (containing over 100 billion nucleotides) and included the complete DNA sequences for more than a dozen species.[7] A whole new field of research called *bioinformatics* has appeared in which biologists try to discover how genes and proteins work by comparing and analysing sequence data.

Big Data and Data Mining are now major areas of research in themselves. To use an analogy with fishing, in the old days scientists would carry out one experiment after another. In effect they were throwing in a line and trying to catch a particular fish. A lot of modern science is more like a fishing trawler. You throw big nets into veritable oceans of data and see what discoveries turn up by serendipity.[8]

Scientists still do experiments, of course, and observe nature, and collect data, but more and more of their data is 'recycled'. Find an insect in a pond, and it might help you understand the ecology of that pond. But combine that observation, the presence of that insect in that pond, with thousands of other observations, and you might make other discoveries that no one could ever make alone.

One use for big data is to make it possible to model complex systems. The most notable success story is weather forecasting. Predicting the weather is such a high priority in so many aspects of modern life that most governments maintain meteorological offices to do the job. Although they use huge supercomputers to turn observations into forecasts, their models would be hopelessly inaccurate without adequate data. To do the job effectively, the scientists need frequent data about conditions across vast areas. It helps enormously if you can trace the path of a hurricane as it rolls in towards you across the Atlantic. For this reason, meteorological offices all over the world share weather information on a daily basis. The World Meteorological Organization (WMO) plays a leading role in coordinating the collection and dissemination of all this data.

Another huge global effort has been the control of disease. This work is coordinated by the World Health Organization (WHO). During the Middle Ages, plagues would periodically ravage the civilized world. They would travel unhindered from town to town, from country to country, spreading misery and death to untold millions. Today, outbreaks are reported at once and the details are passed on to authorities the world over. But the coordination goes well beyond mere reporting. There is active suppression too. One of the greatest success stories was the World Health Organization's eradication of the disease smallpox in 1979.[9]

## The Global Village

The Olympic marathon commemorates the exploits of the Greek soldier Phidippides who, according to legend, ran from Marathon to Athens to warn its citizens of the approaching Persians. During the 1850s, Pony Express riders in the

American west are said to have covered the 3,200 km from Missouri to California in about 10 days, changing horses every 25 km.[10]

When I was a child, making international phone calls was not easy. Undersea cables could handle only a few calls at a time. If you had to make a call, then you contacted an operator to book the details, and some time later they would call back and make the connection for you. You tried to telegraph what you needed to say before the beep, which signalled another 3-min charge being billed.

Nowadays, I can dial up the other side of the world at a moment's notice and talk for an hour at minimal cost. The ability to communicate is making the world a smaller place. All of the world's airlines, all of the world's airports, and all of the world's travel centres are linked by vast communication networks that link together flight schedules, reservations, and departures around the globe.

Advanced communications have made it simple to exchange money anywhere in the world almost instantly. Seizing on this ability, people immediately began to carry out electronic commerce around the globe. Perhaps the most telling results were the establishment of international money markets, and the opening up of stock exchanges around the world to international traders.

The advantages of on-the-spot, up-to-date information are not confined to large organisations. Sue and Ted are a young married couple living in Edmonton, Canada. Thanks to communications satellites, they can watch the Wimbledon final live in full colour, or talk to colleagues in Africa from a mobile phone, or hold a video conversation with members of Ted's family in Australia via the Internet. Meanwhile their friends and family can keep up to date by reading their news blogs on social media such as Facebook.

Sue and Ted also do their business online. They look to investments as a way of boosting their savings and securing their future. They look at prospects not only within Canada, but all around the globe. In the evening they go online and check out the stock exchanges in Hong Kong, Melbourne and Tokyo. In the morning they do the same for London and Bonn. If they find companies they are interested in, they want to find out more. So they might be looking at such widely spread prospects as a tour company based in Dunedin, New Zealand; a chain of microbreweries in Portland, Oregon, or a company building intelligent robots in Edinburgh, Scotland. In each case they access detailed local information, not only about the company, but also about the area, local competition, and so forth. In short, they are able to access detailed information, of many kinds, from all over the world.

When the World Wide Web first became available to the general public, it was seen as a great democratic environment, overturning the dominance of traditional information brokers and allowing individuals a greater voice and power. No longer bound by information filtered by corporate controlled media, individuals could address the entire world through their blogs and home pages. Small businesses could compete with large chains by selling online.

Ultimately however, the sheer scale of the rapidly growing cyberspace provided the opportunity for new power blocs to emerge and control the flow of content. What is the good of placing your information online if people do not know it is there? And even if they know it there, can they find it? This problem led to the appearance of indexes and integrator sites that pooled together links to thousands of web sites with related products, services or information. User preferences, combined with feedback, helped such sites to grow, becoming more influential and powerful as they did so. The process was essentially the same as that which created specialist suburbs such as Tokyo's Electric Town.[11]

At the time of writing, the Internet is increasingly dominated by large commercial resources. Most of the leading companies are household names: *Google* for indexing and searching, *Facebook* for social interaction, *Twitter* for rapid communication, *Amazon* for books, *iTunes* for music, *Wikipedia* for information. Some of the leading organizations have diversified their activities, expanding into new kinds of marketing. Icons for Facebook, Google and Twitter, for instance, now appear on thousands of web pages. The aim is to "lock in" users, to become people's first destination for buying and selling online. Technology assists the locking-in process. Buy any mobile device, such as a tablet pc, and it comes with software that makes web browsing both simple and convenient. But they do so by directing the user to a preferred service provider.

Since its beginnings in the early 1990s, the World Wide Web has grown explosively. When I opened my country's first web server in 1992, there were so few sites in the world, they could be listed on a single menu in Mosaic, the first hypermedia browser. A matter of a few dozen page at most. In 2009 the number of separate domains was estimated to be about 109.5 million.[12] Just 4 years later, the number of indexed pages had increased to nearly 50 billion.[13]

Given such explosive growth, the struggle to index all information on the Web is never ending. A typical search might return a report saying that there are (say) ten million pages. But the report lists only the first ten. No one can sift through ten million pages or to check which are the most relevant and useful. So the search engine tries to help. Using its records of your previous search activity and interests, it provides you with what it decides are the ten most relevant links first. This personalization of web searches ensures that, most of the time, users will find the pages they want near the top of the list.

However, there is a crucial side-effect, a hidden danger, in the way the search engine tries to help. For in selecting what it thinks a user will want to see, it also filters out information that might provide the user with new ideas, or challenge their current point of view.[14] In this way, web searches unwittingly reinforce users' preferences and promote confirmation bias.

Information technology is one of the major forces behind the rapid trend towards globalization. Globalization is the process whereby organizations, institutions and processes expand until they fill the entire globe. By the turn of the Millennium, economic globalization had become an issue that generated division and heated debate all over the world.[15]

Information technology contributes to globalization in several ways. The first and most obvious is communication.

Computer technology is altering the form, nature, and future course of the American economy, increasing the flow of products, creating entirely new products and services, altering the way firms respond to demand, and launching an information highway that is leading to the globalization of product and financial markets.[16]

Governments and corporations are being drawn irresistibly into the global economy. Global business is just not viable unless companies on different sides of the world can talk to each other.

As apologists are quick to point out, globalization is a not a new phenomenon at all. Global expansion of economic and political power has been going on throughout history. What has alarmed people, however, is the speed and scope of globalization, especially its impact on jobs, on people's well-being, and on cultural diversity.

If anyone still clings to the idea that big corporations are more likely to employ the world labor force, to the idea that size begets jobs, here's another shocking statistic from the Institute for Policy Studies: the two hundred largest corporations in the world, and they are getting larger and fewer all the time, account for approximately 30 per cent of global economic activity, but they employ less than 1/2 of 1 per cent of the global work force.[17]

In some spheres, globalization has beneficial side effects.[18] Political global-ization is a trend in which inter-governmental organisations (especially the United Nations) are joining forces to address global issues, such as disease and envi-ronmental management. In public globalization, international public organisations, such as Greenpeace, are forming and growing in power to address issues such as human rights and other concerns of widespread public interest.

One concern raised by the spread of a global culture is the potential loss of local cultural identity. Almost everywhere, traditional cultures are being swamped by western culture. Improvements in communications have played a significant role in creating this threat.

Language is perhaps a good indicator of cultural diversity. A community that speaks the same language is likely to share many cultural ideas and practices. It is estimated that there are about 6,800 spoken languages still in use around the world today.[19] Of these, eleven languages account for 51 % of the world's population, each having over 100 million speakers.[20] Another 200 languages account for 44 % of people world-wide. At the other extreme, speakers of 95 % of the world's languages account for the remaining 5 % of the population. Many of these less common languages are in danger of disappearing altogether.[21] This generally happens when the younger generation adopts other languages and older, native speakers die. This situation is common where a community becomes fragmented and absorbed into a larger cultural group. Although the Internet has contributed to the universal spread of Western culture and languages,[22] it has also provided the means for cultural groups to maintain communication and cohesion even if they become geographically fragmented and isolated.[23]

## Big Brother is Watching

Published in 1948, George Orwell's novel *Nineteen Eighty-Four* told the story of a society in which people's words and actions were constantly monitored by the state. Written as a warning against the repression of totalitarian societies, the scenario seemed very unlikely in mid-Twentieth century Britain. However, by the early Twenty-first century at least one of its premises has become reality.

Almost anywhere you go in public today, your actions are likely to be monitored via closed circuit television (CCTV). The use of CCTV in public places has become almost universal, not to mention many private places too. But unlike Orwell's novel, it is not a case of a totalitarian government seeking to monitor your every word and deed for political control. The use of CCTV by governments is a response to demands to cut rising crime rates. Cameras are used chiefly in public places such as railway stations, car parks, in and around public buildings and centres of nightlife and shopping centres. A study in 2011 estimated that there were at least 1.85 million CCTV cameras operating in Britain.[24] Of these it was estimated that about 33,000 were operated by local authorities, 115,000 were located on public transport and the rest, at least 1.7 million cameras, were operated privately.

In part the spread of CCTV is a cost-effective method of law enforcement. It is simply too expensive to have police officers patrolling every street corner, but you can set cameras to watch instead. The greatest numbers of cameras are operated by private security companies. They are to be found in privately owned properties, especially companies trying to protect their property from thieves, vandals and other threats. The advent of face recognition software makes it possible to track individuals efficiently, especially in crowded areas such as railway stations.

Technology is also being used to improve road safety. Car drivers today are monitored like never before. Ever since motorcars first appeared, and long before, police have patrolled the roads hunting for criminals and watching traffic violations. Today much of this police work is automated. Instead of patrolling the roads, many police officers now man computers. From the late 1990s, automated digital cameras have been used increasingly for traffic control in many cities. They are used especially for catching drivers who are speeding or failing to stop for red lights at intersections. The justification for their use is that they improve safety. On the other hand, critics argue that, rather than a safety measure, traffic fines have come to be seen as an important source of public revenue.

Consider my morning trip from home to the office. Leaving home, I drive down a hill to the main tollway. From there, my route is tracked by cameras for some 15–20 km. When I reach the grounds of my university, CCTV cameras follow my progress into the main car park. They continue to watch as I leave the car park heading for the office. Security cameras on the outside of several buildings record my route and on entering my building, cameras note my every step as I walk through the foyer, head up the stairs and follow the corridor right to my office door. In short, my journey to and from work is monitored the whole way.

Being caught on camera is only the start of the ways we are observed. Advances in communication and information technologies have led to our day-to-day activities being observed in many ways. A side effect of virtually every technology we use to interact with the world around us is an opportunity to gather information about us.

Data about you and your activities have become important commodities to be bought and sold in the information marketplace. Use a credit card to buy new clothes, or pay a bill, and a record is kept about when and where and what you did. Send a text message via your phone and a copy of that message is recorded. Download a file from a website and the server adds a line to its user log.

As an example, take a message sent on Twitter. Everyone knows that Twitter messages are short. What they do not realize is that every message they send, even just a single word, is accompanied by pages of metadata about the sender and the circumstances in which the message was sent.[25]

Many web sites compile profiles about their users. Although most web transactions are supposedly anonymous, servers can keep track of individuals by storing a "cookie" on the client's hard drive. A cookie is simply a file that contains information about a user. The main application of these cookies is to help a server to provide a personalized service to clients. However they also serve to identify the client and enable servers to build detailed profiles about what the client does, their interests, their preferences, and more.

Almost as soon as people started linking computers together, two major problems surfaced as side effects: hacking and computer viruses. At first the perpetrators of these side effects were doing it simply for the rush of proving they could do it. Some viruses were truly nasty, wiping out entire hard drives. Over the years, however, they have mutated into thousands of less virulent, but no less nasty malware species. Instead of destroying the host, spyware sits on your computer gathering data about you that can be used for anything from targeted marketing to identity theft. Protecting computers from such threats has become a multi-billion dollar industry in itself.

Although all of the above data about you are collected by different organizations, the fear is what will happen as governments or corporations begin to pool the information. This is already happening.

On 1st June 2013, journalists from London's Guardian newspaper met with an American defence contractor named Edward Snowden in a hotel room in Kowloon.[26] In the course of interviews lasting seven days, Snowden revealed to the journalists that the United States' National Security Agency (NSA) had been monitoring the emails, Internet use and other communications of millions of American citizens. For instance, it forced the telecom company Verizon to hand over the phone records it held about subscribers. Snowden also revealed that the NSA was using a program called Prism to monitor data held by major Internet companies, including Google, Facebook and Apple. Far from denying the claims, President Obama defended the programmes, saying: "You can't have 100 % security, and also then have 100 % privacy and zero inconvenience".

## The Power of the Media

The power of the media to influence people has long been known and exploited. Companies would not be willing to spend millions of dollars on television and other advertising if it had no impact at all. Needless to say, the nature and extent of media influence has been the subject of intensive research.

The influence of television on people's behaviour has been a hotly debated almost since its invention. Numerous studies have raised concerns about issues such as the long hours that children spend watching television and the effects of the on-screen violence they are exposed to. However, it is difficult to prove the effects of television because it is almost impossible to separate the effects of television from a host of other social factors.

The ideal experiment to resolve this problem would be one in which you could compare a society before and after the introduction of television. Just such an opportunity occurred in 1973, when the Canadian Broadcasting Corporation agreed to install a transmitter to provide reception for a town in British Columbia. Situated in a remote valley, the town had never had television transmission. Tannis MacBeth Williams, from the University of British Columbia, seized the opportunity and carried out a before and after study to observe what happened.[27] She and her colleagues compared social attitudes and behaviour in the town (for which they used the alias "Notel") with two nearby towns of similar size (about 700 inhabitants). One ("Unitel") had a single television channel; in the other ("Multitel") people had access to several channels, as well as access to cable broadcasts.

The results of the study were mostly consistent with other studies into the effects of television. The most pronounced effect was increased aggression amongst children. For example, in the school playground, the frequency of shoving and abuse amongst second grade pupils doubled after the introduction of television. Another large change was the participation in sports events, which fell by half, with the chief decline being amongst older residents.

A surprising result of Williams' study was that the mere presence of television seemed to be what made the most difference, rather than the content of the programmes being broadcast. Another surprising result was that television seemed to hinder the reading ability of kids from poorer households, but had far less effect on those from wealthy families.

Studies such as the above make it clear that television and other media have a significant influence on social issues. That they influence public opinion is beyond dispute. Just look at the vast amounts of time and money spent on television advertising if you need convincing. However, it does raise interesting questions. For instance, can television change consensus within a community? How great is the influence of television (and other media) on public opinion?

My colleagues and I tested these questions by carrying out a simple virtual experiment.[28] First, to approximate the way people live, we assume that a person's daily routine was divided into two phases: day and night (Fig. 17.1). During the *daytime phase*, people interact with one another: they visit friends or go to work

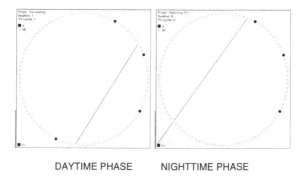

DAYTIME PHASE          NIGHTTIME PHASE

**Fig. 17.1** The dual phases involved in the competition between media and peer influence. In the *Daytime Phase*, the model setup is the same as in Fig. 15.2. Peers influence one another. In the *Nighttime Phase*, people are isolated from one another. However, they are all influenced by the media (shown as a *black dot* at *bottom left*)

and talk to colleagues. During the *night-time phase*, they no longer interact: they stay at home and watch TV. Of course, people's real lives are more varied than this, but remember, it is an approximation.

To test the influence of media, we considered an issue about which there are just two opinions. Call them A and B. The issue could be anything: whether you vote for one political party, whether you agreed with a government decision, or whether you preferred one brand of soap over another. For the purpose of the experiment, we suppose that everyone in the community is in agreement over the issue. They all think A. However, every night the TV pumps out the message that they should think B.

To begin with, consider a scenario in which no one has a TV. Call this *peer-world*. Then in the evenings, there is no TV urging them to change their minds. And during the day, their opinions are reinforced by friends and colleagues. So there would be no change. Now, imagine the opposite: a world in which people stay at home permanently and never talk to anyone else. They just sit at home watching TV all day (such people do exist). Call this *homeworld*. In this scenario, repeated exposure to the TV message convinces more and more people to change their minds. The number of converts increases at a predictable rate, until eventually everyone is converted to B.

However, the more realistic scenario is one where people's lives alternate between the two phases. This is another case of *dual phase evolution*, which we met in earlier chapters. What happens in this scenario is different again.

During each evening phase, the TV converts a few people to B. However, during the daytime phase, these converts are subject to pressure from their peers. At first, most of them get converted back to A. As time goes on, the way in which the media message spreads is completely different. Instead of a steady increase in converts, peer pressure at first suppresses the rate of conversion, keeping it much lower than in homeworld. Eventually, however, the number of converts reaches a

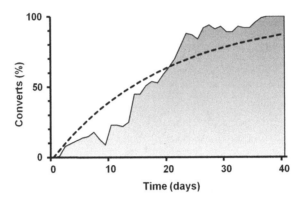

**Fig. 17.2** Results of the simulation experiment on media influence. The graphs show the number of converts to the media's message over a number of days. The *dashed line* is the *home world* scenario (no peer influence) and the *solid line* shows the effect of dual phase evolution in which people alternate between *homeworld* and *peerworld*. Notice that with peer influence, the number of converts is initially slow, but accelerates once a critical point is passed

critical level. At this point there are enough converts that they start to convert others by peer pressure. The result is that the rate of conversions accelerates enormously, becoming much faster than in homeworld. The final result is that the entire community is converted almost twice as much fast as in homeworld (Fig. 17.2).

This experiment teaches us several lessons. One is that, except in the most extreme cases, the media always wins: eventually they convert everyone. Another message is that in promoting any idea or product, it is an advantage to exploit people's social networks. This provides the motivation for viral marketing, in which interest in a product is spread chiefly by social means. And, as further experiments show, the rate of conversion is even faster if the most influential people are converted first.

The ability of the media to manipulate opinions creates the potential to subvert the political system:

> The media have become an alternative establishment, one dedicated to a theatrical distrust of individual politicians and a furious and calculated indifference to the real-life intricacies of policy-making.[29]

The media possess enormous power to set political agendas, to determine what are the acceptable standards of morality or behaviour and to set the context within which public issues are discussed:

> Context is often by far the most important thing about news. ... The same incident can be transformed through rearranging the context: at an extreme, the perpetrator can become a victim, the peacemaker a warmonger, the official striving to tell the truth a bureaucrat disguising the facts.[30]

The problem with the media's ability to set agendas is that the choices they make are often not the agendas that need to be addressed, nor the ones that should be addressed. As we saw in Chap. 5, our animal inheritance includes instinctive behaviours that are honed to ensure survival in the wild. One instinct is to be alert for the unusual, for anything that might herald approaching danger. Another is to be intently alert to any information that bears on social status or other changes within a social group. Translated into modern times, these instincts become tendencies to look for the sensational.

The media's first priority is not the good of society, but attracting public attention and selling their product. And sensation sells. So our news is packed with accidents, crimes, scandals, and political intrigue.

The focus on the sensational tends to trivialize issues. Political debate has become shallower. Instead of parties being evaluated by the quality of their policies; media attention tends to focus on the charisma of the leader.

Implicit in any social question is an underlying assumption about right and wrong, about what is morally correct and what is socially acceptable. The media take the position that they are the voice of social conscience, no matter whether they are acting on false assumptions, misinformation, or special interests.

Reinforcing the above view, the media often claim they strive for "balanced" reporting. In politics, this usually means the news follows up a government statement with one from the opposition party. This approach turns any issue, no matter what its nature, into a political or ideological question. In some cases, "balance" means giving far more credence to extreme views than they deserve.

The trivializing of issues, and the tendency to place them in a social context, leads to a focus on public opinion, rather than facts. Thus policy making is increasingly driven by media reporting and by day-to-day fluctuations in opinion polls.[31] But in modern democratic societies, it is not necessarily the case that the opinions of the majority are a reliable guide as to what policies will be best for the country. That many, or even most, people believe something, does not make it true. The mediaeval world was obsessed with witches, but this did not mean that the poor women condemned for witchcraft were really capable of black magic. For hundreds of years all educated people believed that the sun revolved around the Earth, but that did not make it true.

## End Notes

[1]Marshall McLuhan (1911–1980), from an interview reported in *Playboy Magazine* (March 1969).

[2]CSIRAC (the *Council for Scientific and Industrial Research Automatic Computer)* was built by engineers at Australia's Commonwealth Scientific and Industrial Research Organization (CSIRO) in 1949. It is now on display at the Museum of Victoria in Melbourne.

[3]The HyperText Transfer Protocol (HTTP). This provides a standard way of referring, not just to a computer, but to particular documents or programs stored on a remote computer. It is based on the idea of a client, which sends a request, and a server, which sends back the requested data.

[4]Netcraft (2012).

[5]Actually the mount is closer to $1.18 \times 1,021$ bytes. This is because data volume are calculated in powers of 2, not 10. A *kilobyte* is 1,024 bytes ($1,024 = 210$) and a *megabyte* is 1,024 KB or 1,048,576 bytes. Going up in powers of 1,024, most people are familiar with *gigabytes* and *terabytes*. Large data stores already deal with *petabytes* and even *exabytes*. Then come *zettabytes* and *yottabytes*. Names for larger volumes are not yet fixed at the time of writing this.

[6]Bilofsky and Burks (1988).

[7]NCBI (2007).

[8]Green (2004).

[9]Fenner et al. (1988).

[10]Holzmann and Pehrson (1994).

[11]See Chap. 10.

[12]Wikipedia (2009). http://en.wikipedia.org/wiki/World_Wide_Web

[13]As estimated by WorldWideWebSize.com on 5 April 2013. The number of pages that are not indexed by search engines is probably an order of magnitude greater. This includes huge numbers of ephemeral pages that are generated dynamically from databases.

[14]Pariser (2012) referred to this effect as a *"Filter Bubble"*.

[15]In his book *The Lexus and the Olive Tree*, New York Times foreign correspondent Thomas Friedman describes the process of economic globalization and contrasts the economic forces that driving globalization with the social and historic ties that create an ever stronger sense of place.

[16]US Bureau of Labor Statistics (1998).

[17]Jerry Mander (1999).

[18]Suter (2000).

[19]Throughout this paragraph, the figures given are from UNESCO 2012 (portal.unesco.org).

[20]Mandarin Chinese, English, Spanish, Arabic, Hindi, Portuguese, Bengali, Russian, Japanese, French and German.

[21]See for instance, UNSECO's Languages Atlas. Source: www.unesco.org/culture/languages-atlas/ (downloaded 31 July 2013).

[22]According to *Wikipedia*, in 2009, the English language accounted for 56.4 % of web pages, German 7.7 %, French 5.6 %, and Japanese 4.9 %. See http://en.wikipedia.org/wiki/World_Wide_Web.

[23]See for instance Digital Micronesia (Spennemann 2005) which provides cultural information about the Pacific islands that make up Micronesia.

[24]Reeve (2011).

[25]Cheong et al. (2012).

[26]This story is based on information in Guardian Newspaper's web site *The NSA Files* http://www.theguardian.com/world/the-nsa-files. Downloaded 31 July 2013.

[27]Williams (1986).

[28]Readers can perform this experiment for themselves by linking to Monash University's Virtual Laboratory at http://vlab.infotech.monash.edu.au/. The results are described in several studies, e.g. Stocker et al. (2003).

[29]Lloyd (2004), p. 81.

[30]Ibid.

[31]Megaloganis (2010).

# The Root of All Evil

# 18

*Economists see no limits to growth—ever.*[1]

**Abstract**

The world's economy is based on the assumption of continual growth. Industries have exploited every possible means to sustain and expand markets. One effect has been to increase the levels of personal debt in many western countries. Social fragmentation and isolation help to convert traditional social support activities into service industries. Corporate attitudes and ideas have infiltrated society, creating the consumer society. Values are increasingly measured in monetary terms. Materialist attitudes lead people to value possessions over human relationships and instant gratification over effort and achievement. However, today's global society is reaching limits to its growth.

## The Growth Imperative

As you grow older only the most notable events from childhood remain clear in your memory. One such memory of mine was spending a wet Saturday doing odd jobs to earn enough money for a book I wanted. The store closed at 5 pm, so all afternoon I frantically swept floors and washed dishes, driving my parents to distraction in the process. By 4:30 pm, I had finally gathered the 8 shillings needed. To make sure I reached the store in time, my father drove me there in our car. By 5:15, I arrived back home, the proud owner of my very first book.

Comparing the amounts I spent in 1957 with prices in 2013 reveals how much inflation has increased costs during my lifetime. Today the same book would cost more than 20 dollars. In other words, inflation has increased the price 2,500 %. This means that on average, the price has doubled every 11.8 years since the 1950s.

D. G. Green, *Of Ants and Men*, DOI: 10.1007/978-3-642-55230-4_18,
© Springer-Verlag Berlin Heidelberg 2014

Inflation occurs in every modern economy. In many countries, it is deliberate. Governments create inflation by increasing the supply of printed money. But inflation occurs even when not planned. It is the net result of everyone looking after their own interests, without heed to the consequences elsewhere. Economists have put forth models of inflation that stress different factors. Some, such as John Maynard Keynes, have pointed to a triangle consisting of three factors. First rising prices of key commodities lead to "cost push", in which manufacturers are forced to raise their prices to cover increasing costs. Then there is "demand pull", in which producers raise prices to take advantage of high demand. Finally, there is a wages spiral in which workers seek pay increases to keep ahead of inflation. In short, we could summarize these processes as *need* (cost push), *greed* (demand pull) and *positive feedback* (wages spiral).

Another way to look at this spiral is in terms of time and effort. If you devote effort and resources into producing a product, then naturally you want to get enough return to equal what you put into it. Otherwise, you are losing time and effort. This need drives people to seek the maximum return for their investment. In other words they seek to value their product as highly as possible.

Need also arises in the problem of cost. Putting together any new enterprise is costly. Although it may generate huge returns in the long term, in the short term it can require an extremely large amount of capital to get it off the ground. As we saw in Chap. 7, banks serve to convert an extreme cost now into a mean cost distributed over a longer period of time. But this raises an issue for bankers. Sure they may get their money back in time, but in the meantime they do not have that money on hand. And what about their investment of time and effort in helping their clients? The solution, of course, is interest. So if you borrow money, then you need to earn more to pay back the loan with interest. This immediately requires some growth to achieve.

The amount of money in the world is growing dramatically. In 1990, the value of all goods and services world-wide was $US 21.7 trillion. By 2010, this figure had increased to $US 60.4 trillion.[2] This rapid rise, by a factor of nearly 3 in 20 years, represents an average annual growth of 5 % during the period. Some of the most rapid growth occurred in countries that are rapidly expanding their industrial base, especially China, India, Brazil and the Russian Federation.

The importance of economic growth is that it provides a way to keep income ahead of costs. As such, there is an underlying assumption that growth is not only desirable, but also necessary. The assumption of continual growth is almost literally a dogma amongst economists. Following a meeting to discuss ways for making UK treasury policies more sustainable, a high-ranking treasury official was heard to comment: *"Well, that is all very interesting. Perhaps now we can get back to the real job of growing the economy."*[3]

Because economics and commerce enter virtually every social context, the universal aim of growth is always a factor. It has side effects everywhere. Economic growth is therefore one of the most widespread influences on modern society. In this chapter I will briefly sketch out some of the ways in which the struggle for economic growth has transformed society.

**Table 18.1** Some strategies for increasing profits

*Reduce costs*

| | |
|---|---|
| People | Automation—employ less people |
| | Increase productivity/efficiency |
| | Fewer employees, but longer hours |
| | Replace old employees with young ones on lower salaries |
| | Import cheap migrant labour |
| | Relocate to countries with low labour costs, fewer regulations, less tax |
| Supply chain | Create your own supply chain, including transport and storage |
| | Build expanding empire to get cheap supplies when local sources fail |
| | Use technology to increase productivity |
| Increase efficiency | Quality assurance to reduce losses |
| | Exploit economies of scale—e.g. centralize common functions |
| | Computerization—better tracking of stocks, facilities etc. Just in time processing |
| | Science—new methods |

*Expand markets*

| | |
|---|---|
| Products | Create new products |
| Consumers | Create new consumers—population growth |
| | Create a consumer atmosphere—need for more |
| | Let people run into debt |

## The Struggle to Grow

The demand for constant growth places enormous pressure on the commercial world. The problem in business is that you have to keep making money. But costs keep going up, so you need to make more money. Like Lewis Carroll's Red Queen,[4] you have to run faster and faster just to keep up. Driven by the constant need to grow, businesses have to explore every avenue to survive and prosper. Basically, there are two ways to increase profits: reduce costs or increase revenue (Table 18.1).

In historical times, western economies managed to grow by expanding geographically. They built empires, explored remote parts of the world, finding new products and new sources for existing products. After becoming an independent nation in 1776, the original thirteen colonies that formed the United States of America expanded westward. After reaching the west coast of the continent, they expanded into the Pacific, acquiring colonies and territories amongst the Pacific islands. By 1853 they were able to force an isolated Japan to open its doors to international trade. By 1960 the USA had grown to comprise fifty states, its final additions being the far-flung states of Hawaii and Alaska.

During the last century or more, global corporations have been able to offset rising labour costs in developed countries by shifting production to parts of the world where labour was cheaper. However, most countries around the world are now part of the global marketplace, so the opportunities for geographic expansion are declining. Instead, economic growth is being fuelled in different ways.

Historically, population growth has helped the corporate world to maintain growth. More people mean more markets. However, in most developed nations birth rates are now declining.[5] Some countries (e.g. Germany and Japan) are actually decreasing in population. One answer to this is immigration. In countries such as the USA and Australia, immigration not only fuelled population growth, but also provided a steady source of cheap labour. Population growth has been constant policy not only for industry, but also for governments trying to maintain national economic growth.

In many cities, including my home town of Melbourne, population growth has led to communities that are straining to cope. Housing shortages lead to booming real-estate prices. So people move further and further from the city centre to find affordable housing. As we saw in previous chapters, this urban sprawl in turn leads to choked road networks and long travel times, as people drive further and further to work through peak periods that last much of the day. Providing enough water for everyone is an increasing problem, so water restrictions are common.[6]

Another driver for economic growth is new technology. As we saw previously (Chap. 12), technology is creating new products at an unprecedented rate. It is also helping to automate production and communication systems, which achieve growth by increasing efficiency and decreasing labour costs.

As industry begins to run out of new worlds to exploit, the simple strategy of more of the same no longer applies. In the long run, thinking inside the square is not enough. If a company is to continue to grow, it needs to consider the problem of growth from every possible angle. The side effects of this process have transformed society.

These social changes have not been the result of deliberate planning, but the cascading side effects of technological and other changes. There has been no conscious plan to change society, but changes have emerged as the product of many adaptive changes by companies and individuals. We saw some of these side effects earlier when we looked at the impact of new technologies on the family. The advent of labour-saving devices linked many household jobs to the marketplace. By purchasing a washing machine, dishwasher or vacuum cleaner, you effectively buy out some of the time that you would have spent on those tasks. These devices also created markets for new products, such as dishwashing powder.

In the 1950s, an average family could live comfortably on a single income. You would think that dual income families would be better off, and for a while they were. But the market place absorbed the growth they offered. The greater buying power of two income families meant greater demand for goods and services of all kinds. This demand, together with changing attitudes and aspirations amongst the population, created a spiral of increasing demand and prices. As the cost of living increases, many families still struggle to cope. Today, many families find they need two incomes just to survive.

Changes in family life have served to create many new markets by turning activities that were formerly outside the economy into commercial opportunities. In a traditional extended family, most essential jobs would be carried out by family members. There would always be plenty of adults around to take care of essential household jobs such as cooking, cleaning, gardening, washing clothes, washing dishes, mowing lawns, and taking care of the children.

Perversely, by creating more free time at home, labour-saving devices have left many people with less free time than ever. In a modern nuclear family, or even worse, a single parent family, the adults used up their extra free time by taking on full time jobs. So they have even less free time to do routine household chores than ever before. This means that many unpaid household jobs are now outsourced as paid services.

One such activity is child-minding. In the extended family there are always grandparents, aunts and uncles to help look after children. In the traditional nuclear family, one of a housewife's main tasks was to take care of the children during the day. In a working family, or single-parent family, young children have to be left for hours each day while their parents go to work. This problem led to a huge increase in the numbers of day-care centres.

Another home activity that is increasingly outsourced is food preparation. Tired parents arriving home from a long day on the job have little inclination to spend more of their time preparing an evening meal. They are more than likely to resort to instant meals. This trend led to huge growth in the fast foods industry.

The gods of growth dictate that any product, any transaction, any activity, must be drawn into the financial vortex. In the Third World, destruction of natural habitats has drawn many self-sustaining rural and hunter-gatherer communities into a commercial vortex, where they now find themselves impoverished.

The commercial imperative is also transforming industrialized societies. As Joseph Pearce points out,[7] caring for your grandparents at home is uneconomic.[8] No money changes hands, and it costs very little extra to house and feed them in a loving, caring environment where they are surrounded by their family. But if you put them into an aged care institution, where you pay for someone else to care for them, then money does change hands, so that is economic.

In the above sense, any traditional family activity is uneconomic. Wherever you look in our daily lives, virtually every traditional activity can be replaced by paid services of one kind or another. When family members prepare meals, wash clothes or care for children, no money changes hands. But if you buy takeaway meals, visit a laundromat or dry cleaner, and send the kids to a day care centre, then money does change hands.

Economic and social pressure creates opportunities for commercial activity to creep into every nook and cranny of our lives. Taken to its logical conclusion, our economic system would have a society where every activity, every service for every individual would be a commercial exchange. Friends and families cooperate with one another, so shrink families to individuals. Convert activities where friends get together into leisure industries where people pay to be entertained. In short, isolate people as individuals so their every transaction is economic.

These social changes have not been the result of deliberate planning, but the cascading side effects of technological and other changes in society.

One example that serves to illustrate the moral absurdity of the economic imperative for growth is crime. From a purely economic standpoint, it is possible to make a plausible, but morally bankrupt argument that crime is actually good for society. Yes, in purely economic terms, some kinds of crime can be great for the country.

First, consider the criminals themselves. Those who rob things from others redistribute wealth. Rob a diamond necklace, for example, and you put on the market merchandise that would otherwise rest uselessly around a lady's neck.

Next there is crime prevention. The mere existence of crime forces society to spend money to employ police to safeguard our streets. Individuals buy locks, security alarms, and pay annual fees for insurance and security services. All of these functions create gainful employment for locksmiths, brokers and security personnel.

When someone commits a crime, they generate work for detectives, medical personnel, the media, insurance assessors, and repairmen, not to mention stores and suppliers who may be able to sell replacement items, or repair damaged property.

Next there is the law. When a criminal is caught, there is vast legal machinery to collate evidence, prosecute criminals and to report the proceedings to the public at large. And of course there are armies of people and facilities employed in detaining, guarding and rehabilitating criminals.

Finally, there are the victims. In violent crimes, victims may need medical attention, counselling, and perhaps even on-going protection. Victims of theft will need to buy new goods to replace the ones stolen.

In summary, if viewed in purely economic terms, crime could be interpreted as a channel that creates a great variety of employment opportunities, creates markets, and significantly raises the volume of money in circulation. In short it could be argued that it makes a wonderful contribution to the gross national product. Fortunately, most of us do not view crime in a purely commercial light.

## The Corporate Society

There are two ways to become rich: make more or need less.[9]

The year was 1958. My family was driving home on a Saturday evening after visiting my great grand-mother. We made this trip often and were always treated to take-away food afterwards. Arguments sometimes broke out over where to buy hamburgers. One shop had better ingredients, but the other used real buns, not slices of bread. We had a choice. Every shop was a separate enterprise. Most were run by migrant families. At every one, the hamburgers were unique. You could also order the exact combination of ingredients you liked. Over time we began to alternate between the shops we liked most.

A generation later, the corner shop is under threat. It is being driven out of existence by large chains. Fast food chains thrive on standardization. It is often more efficient to deliver a standard product in a standard way. The customer can take it or leave it. What you gain is a cheaper product. And one that you can be sure will remain the same, whether you are in Portland, Frankfurt or Nagoya. What you lose is the richness and variety. Eventually kids grow up knowing nothing else.

The struggle to grow means that small companies must aim to become large companies. Otherwise they risk being gobbled up by giant corporations. But in the process of growing, a company needs to change, and those changes have important social implications.

Small organizations can work on a personal basis. A single person can know every part of the operation intimately. But in a large organization, no one person can deal with the minutiae of every part of it. Big organizations tend to be impersonal. To ensure that everyone works in a coordinated way, large organizations use rules, standards, procedures, and records. This difference in the way a business is organized leads to changes in attitudes as it grows.

People naturally tend to project their values onto the world around them. And people working in large corporations tend to absorb the attitudes and values necessary for a large corporation to work effectively. So it is not surprising that many aspects of corporate thinking have seeped into society as a whole.

The world today is increasingly corporatized. By the turn of the Millennium, more than half of the 100 largest economies on Earth were corporations.[10] Many observers argue that not only is economic power at stake as a result of globalization, but also our entire social fabric. They argue that modern society has been moulded by the corporate world. As early as 1904, Thorstein Veblen drew attention to corporate domination of contemporary culture.[11] More recently, philosopher John Ralston Saul outlined many subtle ways in which corporate priorities and mindset have taken hold in Western thinking and values.[12] He argues that the corporate world has hijacked modern culture in the interests of continued growth and affluence. Corporate attitudes, values and needs are eating away at the effectiveness of democratic processes.[12] Through the global expansion of powerful multinational corporations, the world is becoming increasingly corporatized and conformist.

We see many symptoms of corporate attitudes in everyday life. One is our preoccupation with time and the general speeding up of Western society.[13] People have become obsessed with time. We are impatient if we have to wait in a queue, or if an elevator takes more than 20 s to arrive. We want instant meals, instant entertainment, and instant service. Anything that delays or requires time and effort is unacceptable.

In the commercial sphere, the "bottom line" is money. So values tend to be interpreted as a single number, in dollar terms.[14] In social terms, the value of a person is seen as their material wealth. Public perception of wealth is based on outward and visible signs: the location and size of their house, the make and model of their car, their clothes and accessories. In large corporations, the value of employees is measured in terms of "performance indicators", and the indicators chosen are variables known to be related to profit. Even in science there has been a clear shift to corporate values in the way researchers are rated. Researchers used to be valued for the quality of research results that mattered. Nowadays, it is more likely to be how much income they have earned in grants and consultancies.

The influence of corporate thinking in modern society extends to the language we use and from there to the ways we perceive and understand the world around us. Saul argues that corporate speak is not meant to inform, but to exclude:

> ... obscurity suggests complexity, which suggests importance.[15]

We can see this transmutation in the way commerce has hijacked certain common words for its own ends. The word *progress* implies movement towards some ultimate goal. In general parlance, it was once taken to mean spiritual or moral advancement, as in "Pilgrim's Progress".[16] However, after two hundred years of industrial activity the term is now equated with commercial innovation and growth.[17] In other words the goal of "progress" is not to become better people but to have and use more material goods.

In similar fashion, in the context of society, the word *development* has acquired the connotation of commercialization. "Underdeveloped countries" are those with pre-industrial economies. Undisturbed forest or land is said to be "undeveloped." The underlying implication is one of worthlessness. Schumacher argued that to say something is uneconomic is to condemn it in the worst possible terms:

> With increasing affluence, economics has moved into the very centre of public concern, and economic performance, economic growth, economic expansion, and so forth have become the abiding interest, if not the obsession, of all modern societies. In the current vocabulary of condemnation there are few words as final and conclusive as the word 'uneconomic'. If an activity has been branded as uneconomic, its right to existence has not only been questioned but energetically denied. Anything that is found to be an impediment to economic growth is a shameful thing, and if people cling to it, they are thought of as either saboteurs or fools.[18]

The word *sustainable* was brought into the environmental debate originally in the sense that we had to adjust human activity so that we could sustain natural ecosystems and species permanently. Somehow along the way the environmental priority originally implicit in the term became inverted. In many quarters it is now commercial jargon for maintaining economic and commercial growth in the face of dwindling resources.

The environmental movement, with its concerns about pollution, has often been seen as a hindrance to business as usual. However, it also opens the prospect for new "environmentally friendly" products. To label products as "green" is now a widespread marketing strategy. Whether the product really is friendly to the environment is another matter entirely.

# A Material World

> That man is rich whose pleasures are the cheapest.[19]

Perisher Valley is a huge ski resort in Australia's alpine region. Each winter, skiers and would-be skiers flock there in thousands. One of the funniest sights is to stand at the top end of the ski lift and watch these wannabe downhill champions arriving at the top of the ski run. Many of them have never put on skis before. Yet they have all the right gear. Down at the bottom of the slope they posture and show off as though they are champions at downhill racing. Their attitude is that if you have the money, if you have the gear, then everything will come instantly. If you look like a champion then you must be a champion. So after posturing and showing off their designer gear, they set off for the slopes. Not for the nursery slopes, mind you, but for the very top of the hill, the expert run, the most challenging, most dangerous slope on the entire mountain. But when they reach the top of the hill, it is a different story. They arrive at the start of the expert run and contemplate what is in store for them on the way down. Their faces turn all manner of colours as reality sets in.

The would-be instant ski champions at Perisher symbolize a transition that commercialism has wrought on western civilization. You see the same attitude in many different guises. Universities see similar attitudes among students: "Why should we have to do any work? We've paid our fees, so just give us the degree." We are not people any more, but consumers.

The Protestant work ethic of the 19th century demanded thrift and restraint. It shunned excess consumption. Today, the message young people get is: "Don't wait, spend your money now and get whatever you want instantly." The concept of instant gratification runs through many aspects of life. Don't try to fix anything; if something goes wrong, just replace it. If your TV breaks down, get a new one. If you don't like your nose, get it replaced. If you don't like your spouse, get a new one. But whatever you do, just spend, spend, spend.

The origins of this rampant consumerism appear to date back to the mid-Twentieth century, and probably much earlier. In 1955, marketing consultant Victor Lebow summed up the commercial viewpoint as follows:

> Our enormously productive economy demands that we make consumption our way of life, that we convert the buying and use of goods into rituals, that we seek our spiritual satisfactions, our ego satisfactions, in consumption. The measure of social status, of social acceptance, of prestige, is now to be found in our consumptive patterns. The very meaning and significance of our lives today is expressed in consumptive terms. The greater the pressures upon the individual to conform to safe and accepted social standards, the more does he tend to express his aspirations and his individuality in terms of what he wears, drives, eats, his home, his car, his pattern of food serving, his hobbies.
>
> … We need things consumed, burned up, worn out, replaced, and discarded at an ever increasing pace. We need to have people eat, drink, dress, ride, live, with ever more complicated and, therefore, constantly more expensive consumption. …[20]

This approach to marketing has been enormously successful. The commercial world now moulds our attitudes and aspirations. In western societies today, people are constantly bombarded by advertising. It has been estimated that Americans are exposed to more than 3,000 items of advertising every day.[21] As Kate Soper put it:

> ...virtually all representations of pleasure and the life we should aspire to come from advertising, with its incessant message that our happiness is dependent on consuming ever more 'stuff'.[22]

In part, this manipulation of whole generations is a result of clever practical application of discoveries about human psychology:

> ... psychological discoveries have more often been used to manipulate consumers, and to increase our consumption through advertising, with its empty promises of sexual fulfilment. ...
>   We have been mesmerized by the ease and enjoyment offered by the technological feats of the last century – electricity, telecommunications, health care, entertainment, rapid transport and domestic conveniences – yet our consumer sophistication has not yet been matched by our psychological maturity or understanding.[23]

Many authors argue that the resulting materialistic society has had devastating effects on our personal lives. For instance, in 2008, the leader of Britain's Conservative Party, David Cameron, described the United Kingdom as

> ... a society characterized by knife crime, poverty, ill health, family breakdown and worklessness ... Children grow up without boundaries, thinking they can do as they please. ... a decades long erosion of responsibility, of social virtue, of self-discipline, respect for others, deferring gratification instead of instant gratification.[24]

Based on long experience as a child psychologist Sue Gerhardt argues that

> Even in the midst of material comfort and physical security that our forebears could only dream of, we continue to act as if we were deprived and must compete with others to get as much as we can. The reason for this may be that although we have relative material abundance, we do not in fact have emotional abundance. Many people are deprived of what really matters. Lacking emotional security, they seek security in material things.[25]

Studies have shown that children isolated from their parents become emotionally numbed by the experience. During the 1940s, for instance, parents in Britain were not allowed to stay with their hospitalized children; contact was limited to a brief visit each Sunday.[26] In the late 1940s, James Robertson carried out a 2-year study of young children who were forced to spend extended periods in hospital for tuberculosis.[27] Robertson's study found that these children went through three stages that resemble the stages of grief described in Chap. 5:

*Denial*—children cried, protested and looked for their missing mothers;
*Despair*—they became listless and despondent;
*Detachment*—children showed little interest in their parents when they visited, only in the sweets and gifts they brought with them.

These studies as well as subsequent research suggest that emotional detachment in childhood may lay down the basis for materialistic attitudes later in life. Research by Ken Sheldon showed that teenagers with the most materialistic attitudes had the most conflict and aggression in their dating relationships, and demonstrated the least empathy and satisfaction in close relationships of all kinds.[28] Kasser argues that the more materialistic young people are, the less satisfying their relationships will be to them.[29] This research also showed that the ability to form relationships was related to the early care children received in their families:

> Individuals who have not had their needs well met in the past come to think that wealth and possessions will bring them happiness and a good life.[29]

## The Limits to Growth

Small is beautiful.[30]

The essential paradox of today's world is that the commercial system needs and demands growth without limits, but ecology imposes finite limits to growth. As we saw above, economic activity has always been based on the premise that growth is necessary. Growth is an ingrained assumption of traditional economic theory. In the following chapter we shall see that in ancient times, many civilizations disappeared because they exhausted their environment. For centuries, Western civilization avoided the problems of achieving sustainable balance between exploitation and conservation simply by expanding. European countries created empires from which they could extract materials that had become depleted at home. In more recent times, international corporations have extended that principle by moving manufacturing and production to Third World countries where they can exploit cheap labour and less stringent working conditions.

In the natural world, resources impose limits on growth. In the commercial world, the limit is the size of the market. What businesses and governments have been doing is to use short-term measures to increase the size of their market. Take the problem of selling a particular product, such as a car or TV set. New products go through several well-recognized stages. First there are early adopters; prices are high and only the rich and adventurous buy the product. Then, as demand picks up, there is a rapid growth phase. Prices fall as demand increases. The growth part of the above curve is exciting, but as the market enters a saturation phase nearly everyone has the product and demand falls off. Finally sales are limited to replacements of worn out items.

So the problem is how do you sell more cars or TVs when everyone has one? Well of course you convince them that they need a second car and a TV in every room. And what do you do when there is a TV in every room? The answer is new technology. You create digital TV that gives a crisper, sharper image on a flat screen. Then you create home theatres with enormous screens, then 3D images. Then you integrate TV and video, then video to your computer, and your phone. The permutations are endless. And of course, in a materialistic society, consumers just have to keep up with the latest and greatest.

Another limit to economic growth is people's buying power. One solution is to have people spend money they don't yet have. In other words, encourage people to borrow money. As discussed earlier (Chap. 7), this has long been a way of coping with extremely high prices by spreading them out over time. The spread of credit cards has made it much easier for people to spend money they do not have, not just for one-off large payments, but as a way of life.

For the United States, census data show that the average annual income in 2009 was $45,000.[31] However the median debt for all households at the time was around $70,000.[32] The biggest components of household debt were home mortgages and vehicle purchases. However, unsecured debt (including credit cards, personal loans) stood at around $7,000.

Many theorists would argue that economic growth can continue indefinitely. Advances in technology and efficiency will continue to allow growth, with economic activity fuelled by human capital. They point to the growth of information as an example of economic activity that is independent of natural resources.

However, the question remains whether a steady-state economy is possible in the long-term. The concern over the limits to growth question has given rise to new fields of research, such as Ecological Economics, which treats economic activity, and society as a whole, as subsystems that sit within the context of the natural environment.[33]

## End Notes

[1]Jackson (2008) was commenting on the assumption of constant growth that economists make.

[2]Based on figures from the World Bank. http://data.worldbank.org/indicator/NY. GDP.MKTP.CD/countries/.

[3]Quote recounted by Jackson (2008).

[4]The Red Queen, a character in Lewis Carroll's story *Alice through the Looking Glass*, had to run faster and faster just to stay in the same spot.

[5]http://data.un.org/

[6]UNESCO (2012). Managing Water under Uncertainty and Risk, UN World Water Development Report 4, Vol. 1. www.unesco.org/new/en/natural-sciences/environment/water/wwap/wwdr/wwdr4-2012/Downloaded 25 July 2013.

[7]Pearce (2001).

[8]Here the term "economic" means that an exchange contributes to the economy. Anything that does not contribute is "uneconomic."

[9]Buddhist saying.

[10]Anderson and Cavanagh (2000).

[11]Veblen (1904).

[12]Ralston Saul (1997).

[13]Gleick (1999).

[14]See Chap. 8, "Playing by the numbers."

[15]Ralston Saul (1997), p. 50.

[16]Paul Bunyan's book *The Pilgrim's Progress* (1678) was an allegorical tale about religious values. It follows the trials of the character Christian during his journey to reach the Celestial City.

[17]Wright (2005).

[18]Schumacher, (1973), p. 27.

[19]Attributed to Henry David Thoreau.

[20]Lebow (1955).

[21]American Academy of Pediatrics (2006).

[22]Soper (2008).

[23]Gerhardt (2010), p. 29.

[24]UK Conservative leader David Cameron described UK in a speech in 2008, cited by Gerhardt (2010), pp 11–12.

[25]Gerhardt (2010), p. 33.

[26]Davies (2010).

[27]Described in Robertson (1958).

[28]Reported in Gerhardt (2010), p. 33.

[29]Kasser (2002), p. 34.

[30]Schumacher (1973), requoted by Pearce (2001).

[31]Vornovytskyy et al. (2012).

[32]United States Census Bureau (2012). Note that I have used the population average for wages, but the median for debt. So the two are not strictly comparable. However, for distributions such as these, which are positively skewed, the mean exceeds the median value. So the average debt would be greater than the figure quoted.

[33]See for instance Costanza et al. (2007), Daly and Farley (2010).

# Fouling the Nest

# 19

*The power of man has grown in every sphere, except over himself.*[1]

**Abstract**

In recent time, economic and population growth has been sustained by removing wilderness, by removing problems to remote areas and by techno-logical advances such as the Green Revolution. However, unconstrained growth is now encountering the physical limits of the natural environment. People fail to respond to environmental change because the vast scales on which the changes occur render them largely invisible on a personal scale. Maintaining complex ecosystems in balance is difficult, and human attempts often lead to further environmental problems. Both natural and human history show that environmental disturbances and degradation can lead to catastrophe.

## A Finite World

South of Canberra, Australia's capital city, there lies a vast wilderness region, known as the Snowy Mountains . Best known for its ski fields in winter, in summer it is a wilderness of steep hills, fast flowing streams and windswept forests. It is also the source of the legendary Snowy River. One summer I spent an entire week trekking through the Snowy Mountains with a group of friends. The highpoint of our trip (literally) was climbing to the summit of Mount Kosciusko, the highest point on the continent. With an elevation of just 2.2 thousand metres, it is a tiny mountain by world standards. Nevertheless, its observation deck did afford us a view of unbroken wilderness stretching to the horizon more than a 160 km away.

D. G. Green, *Of Ants and Men*, DOI: 10.1007/978-3-642-55230-4_19,
© Springer-Verlag Berlin Heidelberg 2014

When confronted with this panoramic view, with nothing but wilderness in sight, it was easy to be awestruck by the vast expanse of the natural world, by our insignificance before nature. Twenty years later I visited the mountain again and the scene was just the same. Vast, eternal, untouchable—or so the wilderness seems. When we see the true size of the world around us, our senses tell us that the world is big, far bigger than any influence that we might have on it. And yet that impression is false. It is biased by the limits of individual perception.

It is difficult to imagine how humans could impact such vast regions. But single events on this scale do happen. In most cases, wildfires today are ignited by people. In February 2009, the Black Saturday bushfires in the Australian state of Victoria, killed 173 people and destroyed more than 2,000 homes. They also burnt a total of 330,000 ha of bushland. One of the most serious fires was started accidentally by faulty power lines.

Between January and March 2003, wildfires in the alpine region of north-eastern Victoria burnt over 1 million hectares of forests, mostly in national parks and other public lands.[2] For an entire month the fires burned virtually unchecked. Fortunately they did not destroy any towns, but as they spread through inaccessible wilderness, authorities were almost helpless to stop them. The fire cast a pall of smoke over the entire eastern half of the state of Victoria and southern New South Wales. While travelling northeast from Melbourne to Wodonga, one moment I was driving through clear air, the next moment I hit a wall of smoke that was like a pea soup fog for the next 100 km. The fires stopped only when cooler weather intervened; in places they had almost reached the sea before they ran out of forest to burn.

Those fires were caused by lightning strikes, but critics blamed pressure from the "anti-burning lobby" for a lack of prescribed burning to reduce fuel loads.[3] Eucalypt forests are adapted to fire and preventing small fires merely allows fuel to build up to disastrous levels. The same problem has been identified in Yellowstone National Park, Wyoming. In September 1988, fires there burnt out 320 thousand hectares.[4] Up until then, the prevailing wisdom had been that all fires should be suppressed immediately. However this policy overlooked the natural role that fires play in such ecosystems. Having learned from the 1988 disaster, park rangers now allow natural fires to burn themselves out, provided they do not occur under extreme conditions or threaten lives and property.

What is not generally appreciated is that in many parts of both North America and Australia, the environment has been shaped by regular fires lit by humans over thousands of years. The principle aim of this firestick farming was to clear away undergrowth and make it easier for hunters to chase game. An unintended side effect of this primitive technology was to favour fire-prone species, helping them to flourish and spread.

So much for the inability of individuals to affect the environment. The world around us may be vast, but it is finite. By lighting a fire, even a single individual can destroy thousands of hectares of forest. By lighting thousands of fires, by clearing thousands of plots for farming, by damming hundreds of rivers, and by pumping gases into the atmosphere, billions of individuals can affect entire continents, and even the entire planet.

# The Problem of Stuff

One of the great delights for visitors to Japan is to encounter the polite formality of their traditional retail practices. Every purchase is a ceremony. In most stores, assistants bow and greet customers with *"Irrashaimasse"* ("welcome") when they enter. Items you buy are neatly wrapped in beautiful paper and handed over in elegant shopping bags. At home I have a collection of Japanese wrapping papers that were so gorgeous I kept them as works of art in themselves.

Such practices have a long tradition. In the late 1800s, several of the great impressionist artists, including Manet, were deeply influenced by Japanese woodblock prints. They discovered these prints because they were used to wrap tea boxes for shipping to Europe.

The down side of all this elegance is that vast amounts of time and effort are spent on producing beautiful wrapping that is simply thrown away by the purchaser. Like many things in life, this is a matter of trade-off. It is a balance between practical economy on the one hand and making the necessities of life bearable and even pleasant on the other.

There is also the issue of convenience. In the 1980s a television advertising campaign promoted the convenience of their disposable product: *"And when you're finished,"* the announcer screamed, *"just throw it away."*

The need to dispose of items arises in part from the question of what to do with all the stuff we acquire. Like stones, people gather moss if they settle too long in one place.

The problem with stuff is that you have to keep it somewhere. In the first house I owned, this was easy. We had few possessions, so we just put things anywhere and the problem was solved. A single bookshelf had plenty of space for half a dozen books. But stuff accumulates. That's the problem. Whether books, clothes, gadgets or even food, if more stuff comes in than goes out, then the volume of stuff grows.

This growth of stuff becomes a problem when you reach the storage limits of your house. For nearly 10 years we lived in a large house. With so much room, we tended to keep stuff rather than throw it out. When we bought a new sofa, the old one moved into the den. Unused toys and sports gear went into the spare room just in case they were ever needed again. Then we moved to a new house in a new city. The new house was only half the size. We sold or gave away large items of furniture. Even so, when we moved into the new house, the dining room was literally full of boxes. It took many sales, many donations and many garbage deposits to sort it out. Even so, the new house was full from the start. So any suggestion of buying anything new is always countered by the questions: "Where do we put it?" and "What do we remove to make room for it?"

The other problem with stuff is the amount of effort required to keep it in order. Ideally there is "a place for everything and everything is in its place."[5] But when people are too busy or too tired, they often leave things where they last used them (just ask any overstretched parent). So things get in a mess and things get "lost."

This is especially true of things that don't fit neatly into the way we organize storage space. And of course, once the volume of stuff exceeds the available storage space, then you are no longer able to store things where they should be anyway. So our ideal model of a "tidy home" fails.

The problem of stuff is most acute when it comes to dealing with waste. In industrial societies, disposing of household waste seems simple to the average person. You put out the rubbish bin each week and someone comes and empties it for you. You flush the toilet and the waste disappears. You empty the washing water and it drains away. In each case out of sight is out of mind. Once the waste goes, we are no longer concerned about it. And yet, something does happen to all that waste. The U.S. Environmental Protection Agency estimated that each year America produces about 250 million metric tons of household waste.[6] If gathered in one place, this would make a pile ten metres high and covering an area of 25 km$^2$. That is enough to bury an entire town of more than 100,000 people.

On a small scale, the domestic problem of managing stuff reflects part of the problem of managing stuff on a global scale. For thousands of years, the world's population, and its need for resources, were far smaller than the volume the world can support. There was plenty of room. For centuries, people have behaved as if the world around us is infinite. Cut down trees and there were always plenty more. Dump effluent into the ocean and it is gone. Release toxic gases into the atmosphere and they are gone. Kill any number of native animals, catch any number of fish and there will still be plenty more where they came from. At least, that was the assumption. The world seems so big that it is easy to fool yourself into thinking that nothing you do can has any impact on it. Trees are so long-lived that to the casual observer, forests appear unchanged over an entire lifetime. All of this gives the false impression that the world around us is stable, unchanging and unaffected by human activity.

Perhaps the greatest problem facing today's global society is the tension between the commercial world, with its demand for constant, unceasing growth, and the natural world, the environment in which we all live. In the natural world, resources impose limits on growth. There is a *carrying capacity* for every population. This is a maximum size limit imposed by food, water, nesting sites and other resource factors. As a population grows, diminishing resources slow its rate of growth until it finds a balance somewhere just below the carrying capacity. On the other hand, if a population grows too quickly, it shoots past the carrying capacity then crashes, leading to cycles of boom and bust.

As we saw in previous chapters, human population has grown geometrically over the past 200 years. The UN estimates that on 31 October 2011, the world population reached 7 billion.[7] Of these, 18 % live in developed countries, while the remaining 82 % live in so-called developing countries. As well as population growth, there is a huge imbalance in the distribution of resources. The 18 % of people who live in developed countries control 80 % of global resources. This leaves the other 82 % of the people to get by on just 20 % of the world's resources, and "many barely survive on less than $2 a day, and often just $1 a day".[8]

For centuries the Western world has been able to avoid confronting the problem of limited resources by grabbing riches from the rest of the world. Instead we have used short term measures to increase the world's carrying capacity. One measure was the Green Revolution,[9] an on-going series of research and development initiatives, chiefly in the decades following World War 2, that increased crop production world-wide. An important focus was new crops, such as high-yielding cereals. Developments also involved improvements in all areas under agriculture, including fertilizers, irrigation, management, pesticides, and mechanized harvesting. The resulting increases in food availability have been credited with saving the lives of over a billion people from starvation. The problem is that food production was treated as a closed box: find enough food to feed people. The problem of unchecked population growth continues in many of the countries most affected.

Another short-term measure that people have taken to cope with population growth is clearing land for settlement and cultivation. In the period 1990–2007, deforestation world-wide totalled 1.4 million square kilometres. More than 450,000 km$^2$ of forest were lost in Brazil alone.[10]

One of the worst-affected areas was the Amazon Basin, a vast region covering some 6 million square kilometres in northern Brazil. Renowned world-wide for its rainforests and biodiversity, the Amazon Basin was a virtually inaccessible wilderness until the middle of the Twentieth century. The richness of the region's rainforests convinced people that the land would be ideal for agriculture. The building of the Brasilia-Bellem road in 1960 opened the region to development. By 1980, more than 20,000 km$^2$ of forest were being cleared each year for farming and the timber industry. Fooled by the lushness of the Amazon rainforests, developers assumed that the land would make bountiful farmland. What they did not realize was that the forests owed their lushness to nutrients residing in the plants, not the soil. As farms failed, the farmers were forced to clear more and more land just to survive.

What happens when you run out of forests to clear, wilderness to develop, and Third World to exploit? Improvements in the efficiency of agricultural practices, and the use of genetically enhanced crops, are only stopgap measures. Sooner or later, growth has to halt. Sooner or later, world economics has to change:

> We in the lucky countries of the West regard our two-century bubble of freedom and affluence as normal and inevitable ... Yet this new order is an anomaly: the opposite of what usually happens as civilizations grow. Our age was bankrolled by the seizing of half a planet, extended by taking over most of the remaining half, and has been maintained by spending down new forms of natural capital, especially fossil fuels.[11]

## The Limits to Growth

In the year 1859, Thomas Austin released 24 rabbits onto his farm in the Australian colony of Victoria. He had imported them from his native England with the aim of making the alien Australian landscape more like home. His idea, shared by many settlers at the time, was to introduce familiar animals that he could hunt for food and sport. The long-term consequences of this plan were never considered.

Released from captivity, the rabbits thrived in their new surroundings and bred quickly. Over the next century, the two dozen original rabbits grew into teeming hordes. Within 10 years, the introduced rabbit population had grown to over 2 million. By the mid 1870s, the rabbits had spread into the colony of New South Wales. At about this time, the colony of South Australia introduced legislation, the Rabbit Destruction Act, to help combat the problem. In the 1880s, a bounty on rabbits in New South Wales backfired because professional rabbit hunters would release rabbits into fresh areas to maintain what had become a very healthy income.

The rabbits destroyed crops and put pressure on the native fauna, with which they competed. It was one of the fastest rates of spread by an introduced mammal the world had ever seen. By the early decades of the Twentieth century, rabbit numbers had grown into hundreds of millions. In some places, such as the Queensland-South Australian border, governments erected fences hundreds of miles long in futile attempts to stop the rabbits' spread.

Not until the introduction of the disease Myxomatosis in the 1960s (and later Calicivirus) were effective ways found to halt the rabbit plague. When I was a young child, rabbit was a staple source of food; today it is a rare and expensive delicacy on the dinner menu.

The Australian rabbit plague is a dramatic example of how an action carried out by a single individual for local self-interest can blow up into a long-term, continental-scale problem. In our dealings with the natural world, we humans have behaved very much like ants. We go about our everyday business without being aware of the large-scale or long-term implications of our actions. The history of humanity's interaction with the natural environment is replete with the kinds of issues we looked at in previous chapters, including flawed models, limited thinking and local self-interest.

As explained in the previous section, the essential problem of the world today is the incompatibility between economics, which insists on growth without limits, and ecology, which imposes limits to growth. Economics insists on interpreting everything in terms of a single number—the monetary value. This includes both renewable resources, such as labour and manufactured goods, as well as nonrenewable resources. If anything, the drive is to minimize the cost of raw materials, as they impact on prices throughout the system. The result of devaluing natural resources is that the world has been living off its capital, rather than seeking sustainable solutions.

The practice of reducing everything to its monetary value poses a huge challenge for environmental conservation. What is the dollar value of a patch of rainforest? If we adopt a purely economic model, then a patch of undisturbed rainforest is uneconomic. It has no value unless money changes hands. In other words, it has to be "developed." If you leave the rainforest in a state of wilderness, then economics regards it as useless, with no right to exist.

To combat the treatment of environment as an item of commerce, environmentalists have tried to turn the debate on its head by valuing economics in ecological terms. One result is the idea of carbon trading. This measures the environmental cost of economic activity in terms of its "carbon footprint"—the amount of carbon the activity releases into the atmosphere. This new approach has helped to shift the balance, so that it is economic development that has to justify its existence, and not conserving wilderness.

A drawback of the above approach is that it tends to reduce discussion of environmental problems to a debate over the concentration of carbon dioxide in the atmosphere. The danger in this debate is that it tries to reduce a complex issue to two questions: are global temperatures increasing? If so, then is human activity to blame? Although the scientific evidence is a clear yes to both questions, the implicit all-or-nothing approach has polarized popular opinion. At the time of writing this, there is furious debate worldwide about climate change. In effect, the entire environmental debate has been reduced to an argument about a handful of numbers, especially average temperature and carbon concentration. The implicit assumption is that if denialists can refute or even cast doubt on the numbers, then they will have shown that people are having no effect on the environment: that there is nothing to worry about, we can go on clearing, wasting and polluting with no consequences. On the other hand, environmentalists are forced to defend this single point at the expense of urgent environmental issues.

The risk is that fixation on a single issue overlooks all the many other ways in which humans affect the environment. As we saw previously (Chap. 13), almost any problem involves competing aims, depending on how you look at it. Focussing solely on any single interest can make it difficult, if not impossible to achieve acceptable results, let alone optimal ones, for other objectives.

On the positive side, decades of environmental campaigning have raised consciousness of environmental issues. Fifty years ago the environment was totally ignored on most political agendas. Now it is actively debated in international forums. This is progress of a sort.

The world is finite. On a regional scale, the expansion of cities has led to tension between the need for space where people can live and the desire to conserve areas of native bushland. These conflicting aims lead to unforeseen problems. The problem with environmental protection is that it is a global problem that so often competes with local priorities. For example, as urban areas expand, local councils are faced with the competing aims of preserving parks and forest while at the same time opening up land for housing. One response has been to attempt a compromise by allowing people to build homes, but not to cut down trees.

In 2002, Liam Sheahan was a resident of Reedy Creek in the Australian state of Victoria. His new home was set amidst scenic bushland, but he became concerned about the potential fire danger to his home posed by the many trees on and around his property. Wanting to make a fire break, he cut down some 200 trees around his home.[12] By removing the trees, he defied a rule that forbade removal of trees in the area. He was prosecuted and the court fined him $50,000. Legal expenses cost Mr. Sheahan a further $50,000. Six years later, all the homes in Reedy Creek were destroyed by the disastrous *Black Saturday* bushfire,[13] all but one: the home of Liam Sheahan. This local incident highlights the problem of trying to satisfy incompatible goals.

Stories about the winners of lotteries show that many people cannot cope with a sudden change in their fortunes. Finding that they no longer need to scrimp and save to survive, many lottery winners lose all restraint and quickly squander the fortune they have won. The reverse is also true. If you have been rich all your life, and suddenly find yourself poor, then it is hard to change and adopt a more frugal lifestyle. This is comparable to the situation facing Western society today. For centuries, we have been living off the planet's capital, freely spending its natural resources, disposing of our waste without regard for the consequences elsewhere. But now we find that resources are finite, that our garbage is piling up on our doorstep and that the fortune in natural resources we inherited is fast running out.

Faced with such unpalatable truths, many people cannot cope. As we saw earlier, they go into denial. Denial based in local self-interest hinders efforts to reduce atmospheric carbon and reduce potential greenhouse effects. Denialists use all of the approaches described previously (Chap. 5) to sow doubt in the public mind.[14] Below are two of the most frequently used arguments.

*Argument #1. Human activity is too small to have any effect on climate.* This argument centres on a claim that natural sources, such as volcanoes, provide the bulk of the annual $CO_2$ output. Denialists argue that $CO_2$ emissions resulting from human activity form only a tiny fraction of the total. The key here is the word "tiny." Atmospheric $CO_2$ emissions from human activity amount to roughly 3 % of emissions from natural sources.[15] At first glance this sounds like a small fraction. But this impression overlooks the reality. What happens is that natural processes, especially fixing of carbon dioxide by plants, eliminate virtually all the natural $CO_2$ output from natural sources. This establishes a natural balance. Even though some 98.5 % of total emissions are absorbed, emissions from human sources boost the total outputs above the maximum that can be absorbed. So in effect, only about half of human $CO_2$ emissions are absorbed. The rest accumulates.[16] Records show that the concentration of $CO_2$ in the atmosphere grew from 278 ppm in 1750 to 390 ppm in 2011, an increase of 40 %.[17]

*Argument #2. Recent changes in global temperatures are not extreme, but merely part of the natural cycle.* This argument reduces to a question of means versus extremes. Opponents of Global Climate Change (GCC) argue that the recent spate of hot years is nothing unusual. The increases, they claim, are well within the range of variations seen over the past 100,000 years. That is, they claim that the mean annual temperature is not changing. However, this is to ignore that those prehistoric changes were not so sudden. Even so, climate change in ancient times did contribute to massive environmental changes, including the spread of modern humans and the extinction of numerous species of plants and animals. Recent changes resulting from human activity have been nearly a hundred times faster. It is likely that many plant and animal species will be unable to adapt fast enough to cope with such rapid change.

Despite all the evidence, denialists argue that the case for human-induced GCC is not proven; therefore we should do nothing until the case has been proven. But this approach is potentially fatal. Hidden behind the argument is the assumption that for as long as we know, the world has been pretty much the same. So why should we believe that catastrophe is imminent? And yet catastrophes do happen.

A well known principle in environmental and medical circles is the so-called *Precautionary Principle*. This states that when you have two alternative theories and insufficient data to choose between them, then you should adopt as your working theory the one with the more disastrous consequences. If human activity has no effect on world climate then it does not matter what we do, everything will be fine. On the other hand, if it does, then the consequences of doing nothing could be human disaster on a vast scale.

Applying the Precautionary Principle in practice is not easy. Attempts to limit $CO_2$ emissions through a carbon trading scheme have met with considerable opposition from entrenched economic and political interests.

A worldwide increase of just a few degrees in average temperature would be disastrous. For example, melting polar ice would raise sea levels and flood densely inhabited coastal areas, as well as drowning several island nations. Marginal agricultural areas would become desert and could lead to famine, starvation and social upheaval. So the precautionary principle would dictate that the sensible thing to do would be to stop greenhouse emissions. An international treaty has been put forward to do just that, but some countries have refused to sign.

The fear with respect to GCC is that by releasing too much excess $CO_2$ and other greenhouse gases into the atmosphere, we may set off a chain of events that cannot be reversed. The ultimate example of a runaway greenhouse effect is provided by the planet Venus. In many respects Venus is a twin planet of Earth, but a runaway greenhouse effect in its atmosphere produced the ultimate hell—an unprotected human on Venus would be simultaneously roasted and crushed to death. Fortunately, Earth's greenhouse problems are not so extreme. Being farther from the sun, we receive only half as much solar radiation. Nevertheless, there is a danger of long-term changes to the environment.

## The Imbalance of Nature

In 1991 a group of eight women and men entered Biosphere 2, a closed habitat in Arizona's Sonoran Desert where they were to spend the next 2 years. This $150 million experiment by NASA had two aims: to test ideas for colonies in outer space and to better understand issues for the future management of the Earth's biosphere itself. The habitat was a closed system, consisting of a series of domes that enclosed living spaces and greenhouses. The experiment showed that maintaining a closed ecosystem in equilibrium is a very complex challenge. Despite careful planning, the carbon dioxide concentration was highly sensitive to changes in the ecology. Ultimately, it proved impossible to maintain atmospheric oxygen at a safe level.

The Biosphere experiment showed just how difficult it is to maintain a sensitive, complex ecosystem in equilibrium. Make one mistake, however slight, and positive feedback makes the system spiral out of control. As we saw earlier (Chap. 13) any complex system will eventually settle into an attractor. But that equilibrium may not be the one you started with. It may not be the one you wanted.

And yet, people assume that we can go on exploiting Earth's biosphere, a far more complex system, without adequate understanding or experimentation. As we saw in Chap. 3, environmental scientists have been warning about the need to change since the 1970s. Despite these warnings, and despite the rapidly growing size of the world's human population, many people still maintain that we can go on exploiting Earth's resources indefinitely and without consequences.

We see this problem again and again in the sad history of human attempts at environmental management. In many cases, unforeseen side effects led to problems that were worse than those they started with.

The story of the Cane Toad provides a sobering lesson about environmental management. The Cane Toad is a huge, ugly beast. Despite its appearance, in the early 1930s it was seen as a god-send for an industry in trouble.[18] Sugar cane had grown into an important industry along Queensland's north east coast. However, crops were being severely damaged by the cane beetle (*Lepidiota frenchi*), putting the entire industry at risk. When scientists looked at the sugar industry in other countries, they noted that the cane beetle was not a problem in Hawaii because of a natural predator, known as the cane toad (*Bufo marinas*). The state government decided to fix its problem by importing Cane Toad to eat the cane beetles.

The plan worked wonderfully well, at first. The toads ate beetles in enormous numbers. The sugar industry thrived. By the 1950s, Australia had become one of the world's largest producers of sugar. As an industrial quick fix, the experiment with the Cane Toad was a huge success.

Unfortunately, the introduction of the Cane Toad was a case of treating a commercial problem as a closed box. As we have seen, side effects are common when solutions attempt to treat problems in isolation and ignore the wider context. With the Cane Toad, the problem was the animal's adaptability. Removed from its native habitat, there were no natural limitations on it.

A Cane Toad will eat just about anything that fits into its mouth. Having eaten their way through the beetle population, the toads began to massacre the small animal populations of the region. Their impact did not stop there. From two glands on each side of the back of its head, the Cane Toad exudes poison as a defence against predation: any animal that tries to eat it dies. With ample supplies of food, and no natural enemies, the Cane Toad began to spread. Today they are a huge environmental menace all across northern Australia, ranging from northern New South Wales to Western Australia. Their populations are still spreading.

# The Problem of Scale

One of my fond memories of Canada was visiting wilderness lakes in summer. If you sit by a lake and regard the forest around it, all is quiet and still. Nothing moves; nothing seems to change. It feels as if it has always been this way, and always will be. And yet, sediment we retrieved from the bottom of these lakes showed that the surrounding forest has undergone enormous changes over the past few thousand years. And before that, there were no forests here at all. The lake itself was scoured out by vast sheets of ice, perhaps a 1,000 m thick.

Seen from the vantage point of our everyday lives, it can be difficult to see that any environmental problems exist at all. The trouble is that environmental change can be very slow. A single tree can live longer than many human lifetimes. Change does happen, but it is normally so slow that a single person cannot detect it. When I walk through forests that cover the hills near my home, they look the same as they did when I first visited them decades ago. Just like the Canadian lakes, it feels as if nothing has ever changed. And yet, towering above the forest canopy are grey skeletons, the skeletons of enormous trees that once grew there: trees that were destroyed by the Black Friday bushfires that engulfed the area a decade before I was born. After growing for hundreds of years, millions of trees were destroyed in an instant.

People have great difficulty coming to terms with extremes. Wrapped in the cocoon of the everyday and the familiar, we see threats in anything outside the norm. As we saw in Chap. 7, society employs many devices to avoid extremes. We also saw that science too has long had trouble understanding, or even accepting extremes: the extremely small, extremely large, extremely slow, extremely rare, extremely new and the extremely complex.

It is no coincidence that some of the most contentious issues, issues where opponents scream out their denial, revolve around extremes. One of these issues is the extremely slow way in which natural selection fashions species out of random genetic change. Another is the slow, relentless way in which large-scale environmental change emerges as a side effect of small-scale activity by billions of individual humans.

The problem with a personal viewpoint is that it blinds us to events that occur on a larger scale of time and space. Moreover we tend to discount the possibility of catastrophes that we have never witnessed.

"It can't happen to me." This is the carefree attitude of many people who have never been exposed to any of the nastier experiences that life can provide. It has been easy for human society to adopt much the same attitude about the natural environment. During the past 50 years, however, scientific evidence has made it increasingly clear that not only do environmental disasters really occur; they can also change the world permanently.

As palaeontologists filled in the fossil record in increasing detail, it became clear that geologic history included long periods of relative stability, each dominated by a different mix of plants and animals. These long periods when life seemed to exist in balance give rise to the well-known geological periods Cambrian, Ordovician, Silurian, Devonian, Carboniferous, Permian, Triassic, Jurassic, Cretaceous, and Tertiary.

Although the species mix changed only slowly during each period of balance, the fossil record showed that these periods are punctuated by boundaries where large numbers of species became extinct.[19] The greatest of these mass extinctions marks the boundary between the Permian and Triassic periods when about 99 % of all species became extinct.

A growing body of evidence shows that most mass extinctions were set off by catastrophic events. The most famous mass extinction of all was the disappearance of the dinosaurs. This occurred at what is known as the K–T boundary 65 million years ago. The K–T boundary is the transition from the Cretaceous period to the Tertiary. In 1972 a team led by Nobel Prize winner Louis Alvarez showed that the K–T boundary was marked in the geologic record by a layer of iridium, which is an indicator of a meteor strike.[20] Subsequent research located the crater left by this meteor near the Yucatan peninsula. This discovery stimulated a world-wide hunt, which has identified remnants of other prehistoric meteor craters around the world. It is now clear that meteor impacts, and other cataclysms, are responsible for punctuating most geological periods.

Meanwhile, interplanetary probes sent back photos of one world after another showing that the surface of every world is pock-marked by meteor strikes. That such cataclysmic events were not confined to ancient times, but still occur today, was shown in dramatic fashion in 1994 when fragments of the comet Shoemaker-Levy smashed into Jupiter.[21]

Other research has shown that not only did global cataclysms change the world in the geologic past, the living world today is sensitive to disasters as well. In the 10–15 thousand years following the last ice age, forests in the Northern Hemisphere also suffered cataclysms. For over a century, it has been known from preserved pollen that instead of changing gradually, forests in the past changed in fits and starts. Periods of relatively constant vegetation, known as pollen zones, sometimes lasted for thousands of years. Changes in forest composition happened suddenly and defined the boundaries between different pollen zones. In northern regions, the zones included stages in the northward migration of trees following the last ice-age. Like geologic periods, these zones are marked by constant forest composition and abrupt changes in species at the boundaries. In a study of post glacial forests in Nova Scotia, I discovered that the forest history displayed punctuated equilibria, just like the geologic record.[22] The sharp boundaries between the pollen zones were triggered by major fires. Fires ravaged the existing forests making room for migrating populations to expand into new territory.

Environmental disasters have almost certainly affected human history. A minor example occurred in 2010. The eruption of the volcano Eyjafjallajökull in Iceland released a plume of ash that drifted across northern Europe. This plume disrupted air traffic over northern Europe for weeks, causing widespread economic and social chaos.

The explosion of Thera around 1600 B.C. destroyed Minoan civilization and may have affected events in Greece, Egypt and other surrounding countries.[23] David Keys has argued that a massive eruption of Krakatoa in the year 535 AD disrupted climate worldwide for a number of years and thereby triggered major events in human history.[24] These events included the fall of the Roman Empire and the first recorded appearance of plague in Europe.

## The Lesson of History

Let one person cut down one tree and it does not make much difference to the forest. Let millions of people, over hundreds of years, cut down billions of trees, and the story is very different. In the 1600s, Europe encountered a fuel crisis. So much forest had been cleared for farming that there was no longer enough wood left to supply all the fuel needed by the growing human population. The demands of increasing population outstripped the capacity of shrinking forests to provide an adequate supply of wood. To cope with this fuel problem, communities turned from a renewable resource to a non-renewable one—coal.

Elsewhere in the world, growing populations led to the same problem. Japan and China, for instance, faced fuel crises at roughly the same time as Europe, but found different solutions. Japan coped by enforcing careful forest management. One solution adopted in China was to adapt the way people prepared meals. Slow simmering stews had been a mainstay of family meals, but they used a lot of fuel. The invention of the wok made it possible to cook stir fry, which was fast and used much less fuel.

The lesson of history is that civilizations have never learned to live in harmony with their environment. There is abundant historical evidence to show that disregard for the environment leads to human disaster. The current furore over climate change is reminiscent of events within many doomed societies. Archaeological research shows that many, perhaps most, ancient civilisations destroyed themselves by degrading their environment.[25] In most cases this did not mean that people all suddenly died of hunger or were consumed by storms. Instead, by depleting natural resources, they eventually reached a point where they could no longer maintain their civil and military infrastructure adequately. At that point, they became vulnerable to invaders who wiped them out. This pattern seems to have been the case for the Mayans, for Angkor and for Sumeria.

The case of ancient Sumeria is particularly revealing.[26] Ancient Sumerian civilization consisted of city states located within what is now Iraq. The first of these city states appeared about 4500 B.C. They were some of the first cities to practice intensive, year-round agriculture.

The environmental crisis faced by the Sumerians developed very slowly, over a period of about a 1,000 years. Soil salinity was a major problem. The build up of salts affected harvests. Eventually salinity forced the Sumerians to phase out wheat and replace it with barley. However, the changes were so slow that no change was evident within any single life-time.

Even at the end, the Sumerians were unable to adapt their social structure to deal with their environmental problems. Ultimately, invaders wiped out their civilisation. Their very existence was forgotten until archaeologists discovered the remains of their cities in the mid-Nineteenth century, more than 4,000 years later.

Throughout history, people have always craved "the easy life": safety, plenty of food, comfortable surroundings and enjoyment for minimum effort. Society seems to have a way of reorganizing itself according to its capacity to fulfil these cravings. The emergence of kingdoms created enough spare resources for a few individuals (the nobility) to live in ease and comfort, at the expense of misery in the lives of millions of peasants. The Industrial Revolution made it possible for much larger numbers to live, if not in luxury, then at least in greater comfort, leading to the burgeoning of the middle classes. However, the cost of the affluence currently experienced in Western countries has been bought at the expense of dwindling resources and misery in dozens of Third World countries.

Learning to manage environmental complexity may just be crucial to our long-term survival as a species. As the Third Millennium opens, one of the world's greatest challenges is to find a balance between exploitation and conservation . Finding a balance will not be easy. For over a generation now, environmentalists have been shouting warnings about the looming crisis. The tragic lesson of history is that, time and again, well-established cultures and societies have perished because they failed to adapt to environmental crises. Today, there are no new lands to conquer. There are no new resources to exploit. What we have is all we'll get. As resources shrink, the situation becomes increasingly complex. It is complex because everything people do now affects the entire system. You cannot dig mines without displacing farmers; you cannot create new pastures without removing native forest. There is no such thing as wilderness any more. Effects of human activity are felt everywhere.

## End Notes

[1] Attributed to Winston Churchill.

[2] Wareing and Flinn (2003). *The Victorian Alpine Fires January–March 2003*. The State of Victoria, Department of Sustainability and Environment, Melbourne. ISBN 1 74106 624 7, 222 pp.

[3] Paxinos (2003). States accused over fire risks in national parks. *The Age Newspaper*, Melbourne. March 12, 2003.

[4] Anonyomous (2012). Yellowstone fires of 1988. *Wikipedia*. http://en.wikipedia. org/wiki/Yellowstone_fires_of_1988.

[5] This proverb is variously associated with Samuel Smiles, Mrs. Isabella Beeton and Benjamin Franklin. *The Oxford Book of Quotations* dates it from the 17th century.

[6] Taylor and Morrissey (2004). The figures given here for the size of the mound of waster produced *per annum* are based on the assumption that accumulated waste

has roughly the same density as water. The area calculation is based on families of four each occupying a property of 100 square metres.

[7]UNFPA (2011). *State of world population 2011*. http://foweb.unfpa.org/ SWP2011/reports/EN-SWOP2011-FINAL.pdf.

[8]Source: World Bank. http://data.worldbank.org/.

[9]See Jain (2010) for a history and overview.

[10]Source: United Nations Statistical Data on Forest area, data.un.org.

[11]Wright (2005), p. 117.

[12]Australian Senate debates. Monday, 23 November 2009, Questions without Notice. Natural Disaster Management.

[13]Black Saturday. www.blacksaturday.com.au/.

[14]Washington and Cook (2011).

[15]Source: U.S. Global Change Information Office. www.gcrio.org/ipcc/ (Downloaded 20/12/2013).

[16]See Ballantyne et al. (2012). For recent data about increasing concentrations of $CO_2$ see ESRL (2013).

[17]NOAA Global Carbon Project 2012. See www.climate.gov/ (Downloaded 20/12/2013).

[18]Details provided here are based on Lever (2001).

[19]Eldredge and Gould (1972).

[20]Alvarez et al. (1980).

[21]Levy et al. (1995).

[22]Green (1982), also described in Green et al. (2006a).

[23]Grattan (2006).

[24]Keys (2000).

[25]See for example Ponting (1991), Diamond (2005).

[26]The British diplomat and archaeologist John George Taylor discovered the city of Ur in 1853. Crawford, H. E. W. (2004). *Sumer and the Sumerians*. Cambridge University Press, Cambridge.

# Shaping the Future

# 20

*If I could have made this enough of a book it would have had everything in it.*[1]

**Abstract**

Human self-interest, together with side effects of new technologies will ensure that unexpected trends continue to emerge in the future. The information revolution is still underway. Many social effects of new technologies, especially in automation and robotics, are still to come. Trends in biosciences suggest that a biotechnology revolution may be just around the corner. Modern scientific research is both vast and technical. This drives scientists deep into specializations and makes it impossible for lay people to understand research and its implications. Complex decisions often require tradeoffs, but local self-interest leads people to favour short-term, self-serving solutions over long-term, solutions that benefit all.

## Visions of the Future

When I was growing up, almost every vision of the future portrayed it in glowing terms: flying cars, magnificent buildings, and smiling faces on the people. Implicit in all of these pictures was the assumption that technological progress would automatically lead to a healthier, happier society: a better society. And, in many respects, this assumption is correct. Technology has improved the lot of many people. However, as we saw in earlier chapters, rosy visions are not necessarily true. Technological progress does not always equal greater happiness.

For the most part, science fiction explores possible futures that could arise from advances in science and technology. The best science fiction goes beyond the science and explores the social implications. A notable example was Aldous

D. G. Green, *Of Ants and Men*, DOI: 10.1007/978-3-642-55230-4_20,
© Springer-Verlag Berlin Heidelberg 2014

Huxley's novel, *Brave New World*, which explored ways in which advances in biotechnology might transform society. His depiction of the effects that mind altering drugs and birth control pills might have on social mores was shocking in 1932, when the novel was published, but within 50 years some of its scenarios had proved astonishingly accurate. The technology needed to underpin the story's genetically based caste system does not yet exist, but it is not far off.

Sci-Fi stories are replete with examples of new technologies that have devastating side effects. In the 1956 MGM movie *Forbidden Planet*, for instance, visitors to a distant world find the remains of an advanced, ancient civilization called the Krell. Having succeeded in creating the ultimate technology, the Krell were wiped out in a single night because of the technology's fatal, but unforeseen side effect. Closer to home, in Isaac Asimov's anthology *I, Robot*, the humans build robots with three laws to ensure that human safety is their primary goal. Ultimately, however, the laws backfire because the only way to guarantee people's safety is to imprison humans and prevent them from doing anything remotely dangerous.

In the real world, the side effects of new technologies might not be quite as extreme as in Sci-Fi, but as we have seen, they can transform society. The scenario of *I, Robot* may seem far-fetched, but it is already taking place, even without the robots. As we have seen, the nanny state[2] attempts to ensure public safety by imposing restrictions on any activity that may be harmful. But when taken too far, it limits people's freedom and constricts the quality of life.

Unlike us, ants do not have a vision of the future. They do not plan ahead. They cannot plan because they do not have brains. They simply do what they know, and the ant colony emerges out of their activity.[3] We humans, on the other hand, do have brains. We do plan ahead. And yet, despite all our plans, things still go wrong. One reason for this is the subjective nature of our thinking. We usually plan and act within a narrow context, ignoring wider implications. We tend to put our own self-interest first, to consider only local issues, and we often fail to consider the wider implications and side effects.

This book has identified the sources of many unplanned social trends. Identifying these sources helps us to understand the nature of social complexity and its implications. We began by looking at some of the ways in which complexity leads to the unexpected, and the reasons why people fail to deal with it. We then looked at several ways of understanding social complexity. One of these, the network model of complexity, is important because interactions between people and between organizations form complex networks. Often it is interactions within those networks, such as positive feedback, that produce social trends.

The tendency to think inside the square, to consider every issue as a closed box, is the source of so many problems. The most common example is Merton's "imperious immediacy of self-interest".[4] As we have seen, this takes on many forms. Self-interest outweighs group-interest. The immediate drives out the important. Short-term solutions are favoured ahead of long-term ones. Local issues get placed above global issues.

The desire for simplicity really tells against us. We want things to be simple, so we look for simple solutions to problems. Ralston Saul argues that this is one of society's greatest problems: the tendency to look for a single, simple solution that will solve everything.[5] As we have seen, simple models almost inevitably mean that we fail to see potential side effects. Simple solutions, adopted within a limited context, set off unplanned consequences in other contexts. Then processes such as positive feedback come into play, turning small, local anomalies into large-scale patterns and trends.

## A State of Mind

Five hundred years ago, if someone said they had seen a light moving across the sky, it would have been interpreted as a supernatural portent. Today, it would most likely be put down to a flying aircraft, or some other recent technology. In mediaeval times, odd behaviour was liable to be attributed to witchcraft. Today it might be interpreted as mental disturbance. We like to think that we are not as ignorant as people centuries ago. We like to think that today we are better informed; that we act on rationale grounds. In some degree this is true. People certainly are better educated. And with modern media people are better informed. However, society as a whole is certainly not as well-informed and rationale as most of us would like to believe.

One of the reasons stems from the way science has developed. By "science" here, I mean science and technology in the broadest sense. Paradoxically, the more science learns and explains about nature, the less people understand the explanations. In the mid 1800s, any well-educated person could understand most of the major scientific issues of the day. With a bit of background reading, they could keep up to date with the latest ideas and discoveries in a number of fields.

The rapid growth of science during the 20th century has had a number of cascading side effects. The volume of published research has exploded. A recent inventory listed more than 22,000 scientific journals.[6] These journals publish hundreds of thousands of research papers every year.

The sheer volume of research makes it difficult even for scientists to keep up with the latest results in their own fields of research. At the same time, science has become increasingly technical. This makes it next to impossible for the general public to gain anything more than a superficial understanding of most issues. Even trained scientists find it hard to understand technical details in another field. Their chief defence has been to become increasingly specialized, knowing more and more about less and less.[7]

This need to specialize led to further side effects. One is that most scientists get too close to their subject. They become so engrossed in complex technical issues, they lose track of the bigger picture. They can no longer see the wood for the trees. They know all the minutiae of their subject, but do not see its wider implications. And this failure includes any possible side effects. The scientists at Los Alamos were so consumed about whether they *could* build an atomic bomb; it was not until they succeeded that they began to ask whether they *should* build one.

There is feedback at work here. Failure to understand the bigger issues goes hand in hand with the nature of science itself. The strength of science is that its findings are tested rigorously. Any scientist knows that her work will be scrutinized and tested thoroughly. Scientists know that colleagues will slam their work mercilessly for any mistake or omission. So they are careful to justify every word they say. This makes scientists reluctant to step outside the specialized field where they know the issues and what has been published.

One effect has been to push science into ever-smaller silos. In many places, scientists do not even understand the work of colleagues in the next office. This lack of internal communication retards progress on problems that span different disciplines. In today's world, this includes almost any major problem. The usual solution is to form inter-disciplinary teams. But the first problem the members of such teams have to cope with is how to communicate with each other.

The combination of specialization and reluctance also helps to explain why most scientists are not ready, not willing, and not even able to explain their work to a general audience. This gap contributes to public ignorance about scientific issues. Of course, the untrained public is unable to read highly technical research papers, so they do not even hear about many scientific advances.

But the problem runs deeper. Because the technical language of science is so hard to understand, most people give up trying. Few people with scientific training become teachers, so those pressed into teaching science are likely to have a limited understanding of science themselves. The result is a population in which many people are ignorant of science, and unable to understand science.[8] Worse still, many are hostile to science.

Since people cannot understand science, because scientists do not explain their work, and because they want simple answers to life's problems, people turn to other sources for explanations, such as New Age fads.

Being untrained in scientific thinking, non-scientists do not know how to question arguments and sift evidence. Not understanding the evidence, it is difficult for most people to discriminate between truth and fiction. Unscrupulous "experts" exploit this. The more dogmatic and confident "experts" appear to be, the more likely they are to be believed, no matter how wrong they are (see Chap. 11). I have on my desk a copy of a "journal" that appears in my mailbox every few months from an unknown source overseas. This publication gives every appearance of being a scientific journal. It has the same format as a scientific journal and is full of articles about technical sounding issues. And yet, closer inspection shows that it is pseudo-science, a subtle mouthpiece for a right-wing religious group intent on spreading its worldview.

In a world where rapid advances in technology are driving social change faster and faster, failure to understand technology is a serious problem. In most western democracies, few political leaders have scientific training. Fewer still are active scientists or engineers themselves. So they do not understand technical issues.[9] This means that important decisions involving the impact of highly technical matters are left in the hands of people who do not understand them.

People want leaders who know the answers to their problems. Leaders who appear confident about their course of action inspire trust. So politicians cannot be indecisive. Their policies tend to be driven by ideology, not facts. They ignore scientific evidence unless it supports their position. They try to avoid appearing wrong. Like a horoscope that is vague but sounds positive, politicians try to sound presidential but avoid saying anything definite. Their priorities tend to be whatever will help them achieve power. In democratic countries, this means they prefer short-term fixes that make them popular instead of tackling long-term needs that may make them unpopular with voters in the short-term.

With notable exceptions, major sources of public information usually fail to explain science adequately. For a start, few journalists have scientific training. And, as mentioned above, very few scientists engage in public debate. Even the journalistic ethic of "balanced reporting" can be a problem. The idea is that when reporting an issue, journalists try to take a neutral position. So they present both sides of an issue to provide the audience with a balanced, unbiased account. This approach may work well for, say, political debates, but in some cases it gives unintended prominence to extreme ideas. As we saw in Chap. 5, denialists exploit this to create doubt in the face of overwhelming evidence on issues they oppose.

The media exacerbate the above problem by leaping on political errors and creating mistrust. Opinion polls today are both frequent and influential. Some observers have complained that government policy is like a weather vane, shifting in response to the latest opinion poll.[10] New media have made the problem worse, with constant online postings on every issue.

Inability to discriminate reliable evidence from outright lies is made even more dangerous by the Internet. Pose any question you like to an Internet search engine and it is likely come back saying there are millions of hits. Only the first ten or so of these will be listed immediately. Few users are willing to take time trawling through dozens of sources, let alone millions! Search engines are well aware of the problem, so they rank the outputs. Google, for instance, originally ranked sites according to the number of links pointing to each site. But that ranking may reflect popularity, not relevance. More recently, Google has overlaid all that with rankings that reflect your profile. It builds this profile by recording all of the searches you make. It then selects sites based on that profile. In other words it biases the search in terms of what it thinks you want to see. As we saw in Chap. 17, this leads to the "filter bubble".[11] So rather than receiving an authoritative, or even a balanced view of an issue, you get a picture that reinforces your views. In this way it can promote confirmation bias.

## Priority or Panacea?

Even if we do manage to account for everything that affects a decision, there can still be problems. One of the most common is the limitation imposed by finite resources. This leads to the need for trade-offs.

Suppose you go shopping in Ginza, a famous shopping district in Tokyo. You have 20,000 yen to spend. You could spend all your money on clothes and gifts, but then you would have nothing to buy something to eat. Alternatively you could have a meal at a restaurant, but that would exhaust most of your shopping money. This problem requires a trade-off in the way you spend your money.

Suppose you have a day off work. What do you do? The garden needs work, but if you spend all day in the garden, there will not be enough time to visit your friends. Alternatively, you could spend the day visiting your friends, but then there would be no time to work in the garden. So the balance between gardening and visiting demands a trade-off in time.

Trade-offs are needed whenever there are alternative ways to use the same finite resource. They are common in the living world. For instance, plants and animals have only finite resources to invest in reproduction. To ensure that they have offspring to carry on their genes, there are two basic alternative strategies they can employ. One is to produce as many offspring as they can and rely on chance to ensure that some of them survive. For instance, some insects, such as locusts, breed prolifically, but their offspring are left to fend for themselves. The alternative is to produce less offspring, but devote resources to maximize the chance that each offspring survives. Most birds and mammals nurture their young, thus improving their chances of survival.

Governments face the problem of trade-offs all the time. How do you allocate funds from a finite budget to address all manner of competing needs and demands? Many issues faced by society involve trade-offs. One of the most contentious is the way land is used. On the one hand, there is a need for areas of wilderness and nature parks to conserve the world's biodiversity. On the other hand, agriculture, housing and industry need land to satisfy socio-economic demands. And in rapidly growing urban areas, industry and housing will have competing needs.

Another trade-off is between taxation and services. The more funds a government has, the more they can do by way of community services and infrastructure. However, increasing revenue generally means increasing taxes, which is always unpopular.

These sorts of issues lead to problems of optimization. That is, we try to find the best solution to a given problem. The case in which several different variables are involved is known as *multi-objective optimization*. This kind of problem often involves constraints. To take a simple example, if a company increases its advertising budget, then it would sell more products. However, more money for advertising means less money for production, so beyond a certain limit it will not be possible to keep up with demand.

In human systems, one of the most common trade-offs lies between control and flexibility. On the one hand there is the desire to prevent problems by controlling a system. On the other hand there is the desire for the system to achieve its maximum potential. These desires are often at loggerheads.

As we have seen before, the nanny state is one example. It arises when the concern to avoid dangers overshadows the desire for everything else. One manifestation is excessive legislation (and litigation). Although intended to prevent harm, the side effect is to inhibit social interaction, as in the case of cancelled street parties and other community events.

In raising children, parents face the competing needs to protect children from harm on the one hand and allowing them to grow and learn on the other. Recently, several schools near where I live prohibited children from playing chasing games on school grounds. They were motivated by concerns that a child hurt while playing could lead parents to raise questions about duty of care. In the quest to eliminate accidents, the rule limits children from healthy exercise. It also reduces the opportunities for children to learn an important lesson about self-care and responsibility that falling over once in a while would teach them. By focussing on eliminate the dangers of the extreme, the prohibition also eliminated the benefits of the mean.

Large organizations face trade-offs between productivity and creativity in their workforce. Micro-manage and you ensure that things get done when you want and the way you want. On the other hand, you also stifle creativity and motivation in the workforce, so missing opportunities for ideas that could lead to new products or better ways of doing things.[12]

## The Ghost in the Machine

Being wise after the fact helps you avoid future strife. Even better is being wise before the fact: to avoid problems before they arise. If side effects of new technologies are likely to be the most common sources of unplanned social trends, then what are these new technologies? And what are the side effects that will shape society? Among the millions of new inventions appearing every year, there are probably a few that will transform society. I would not claim to know which particular inventions will prove most influential. As we saw earlier (Chap. 11), the predictions of experts are usually poor. On the other hand, it is clear that some rapidly growing areas of technology are already delivering inventions with widespread side effects.

One example is biotechnology. As defined by the Convention on Biological Diversity,[13]

"Biotechnology" means any technological application that uses biological systems, living organisms, or derivatives thereof, to make or modify products or processes for specific use.

Biotechnology includes a wide range of new methods, such as cloning, but perhaps the most far-reaching consequences come from the study of genes and proteins. In 1990, the United States government launched the Human Genome Project (HGP), a bold initiative to map the entire human genome.[14] For the life sciences HGP was a parallel to such grand projects as building the first atomic bomb, or sending the first man to the Moon. The HGP was expected to take decades to complete. Instead, new technologies for sequencing DNA automatically made it possible to complete a draft map of the entire human genome by the year 2000 and a final map by 2003.

The results of the project were surprising. The 46 chromosomes that contain the human genetic code contain about 3.3 billion base pairs. However, there are only some 21,000 genes in all, a much smaller number than expected, and the DNA code for those genes make up only 2 % of the entire genome.[14] So one of the challenges now facing biotechnology is to understand networks of genes as well as the role of all that non-coding DNA.[15] Uncovering all these complex processes is likely to be an ongoing challenge for science in the 21st century.

Among the early triumphs of biotechnology were discovering how to make many copies of particular genes[16] and how to transfer DNA into bacteria.[17] These technologies enabled scientists to trick bacteria into producing useful substances that had previously been rare. One example was to manufacture interferon, a key element in the body's natural defences, with obvious medical benefits. The ability to identify the amino acid sequences that make up proteins has made it possible to design drugs that target specific diseases. However, as scientists learn how to manipulate the DNA code directly, even greater successes will come.

Imagine being able to change a baby's genes at will. If you could cut and paste in code into the genome of an embryo then virtually anything is possible. For a start, you could determine exactly what genetic characteristics a baby would have. You could remove genes associated with undesirable traits, such as predisposition to fatal diseases. You could also insert selected genes from a great athlete or a famous singer. But the possibilities extend much further.

As we learn more about the way genes regulate growth and development, it may become possible to design entirely new organisms. For instance, in 2011, Farren Isaacs and his colleagues at Yale University succeeded in altering the DNA of *E. coli*.[18] In the DNA code, the codons (triplets) TAG and TAA act as *stop codons*. That is, they mark the end of the code for a particular gene. The Yale team replaced all 314 occurrences of TAG in *E. coli* with TAA. Although this change made no practical difference to the genetic makeup of the organism, it did demonstrate a technology capable of making targeted changes in the DNA code.[19]

Breakthroughs in biotechnology research have already provided many benefits, especially in medicine. As side effects, such breakthroughs also raise legal and ethical questions. For instance, to stimulate growth of biotech industries, the US government decided to allow companies to patent gene sequences. However, it also opened an ethical and legal minefield: companies could take an organism, determine its DNA sequence, and patent its genes. This allowed them to investigate potential products without competition. However, a Supreme Court decision in June 2013 called a halt to patents on naturally occurring genes.

## To Really Foul Things Up

In the 1950s and 1960s, there was much excitement, as well as concern, about a new field of computing called *Artificial Intelligence* (AI). Within a few years, it seemed, science would produce robots and other thinking machines that would be virtually indistinguishable from humans. The reality proved disappointing. Fifty years on and critics have been scathing at the failure of the entire field of AI to make good its claims. As we saw in Chap. 5, the problem is compounded because it is so hard to pin down exactly what defines intelligence.

Although reproducing human intelligence still seems to be a long way off, the field of AI has expanded in scope to become one of the most prolific areas of computing research.[20] Almost every computer science department in the world has an AI research project of some kind. And yet, after fifty or more years of work, thousands of AI researchers have so far failed to produce an artificial brain. This failure speaks volumes about what a difficult challenge it is. On the other hand, this army of researchers has produced a wealth of useful products.

Taken in its broadest sense, AI includes a host of fields that are also inspired by finding computing analogies with nature. Examples include artificial life, genetic algorithms, swarm intelligence, neural networks, and cellular automata, to mention just few. For many institutions, these areas encompass nearly half of all research in computer science. The results may not yet include an artificial human, but hundreds of small advances do add up to a lot.

AI has contributed to many new technologies. Expert systems, for instance are programmes that use collections of rules to emulate the knowledge of experts in specialized topics. These systems play important roles in e-commerce and data mining. They have also made possible the increasingly sophisticated forms of automation now built into many products and systems.

Arguably the application of AI with the greatest range of social side effects is automation. Writing for Forbes Magazine, Christopher Steiner warned that

> ... the technology and theory behind algorithmic trading is now entering all sorts of other fields, from advertising to healthcare to pop music creation. There are even algorithms being used on you and me right now that listen to us speak, decide what kind of person we are and then decide how company X should treat us in the future.[21]

It is worth taking a closer look at one area of automation—robotics. Robots used to be big clumsy devices with thick wires joining them to a computer that controlled all their operations. Improvements in miniaturization made it possible to localize the processing required. This means that instead of being referred back to a central processor, movement and other functions can be controlled locally. This ability has led to robots that are smaller, self-contained and cheaper.

Like computers themselves, robots are mutating into many different forms and functions. The field of *swarm intelligence* is borrowing lessons from nature to learn how to make swarms of mobile robots to share information and coordinate their behaviour. At the time of writing, robot soccer has become a highly competitive sport.[22] It provides a practical task in which different research teams can test their models for robot collaboration against one another.

Current surveillance methods rely on relatively small numbers of fixed cameras. In the future these could be replaced by swarms of mobile robots. Like ants searching for food, or even flying like bees, robots could actively seek out incidents, trail individuals, or search areas more thoroughly than any human team could do.

Miniaturization is an industrial goal that continually spawns new technologies. For instance, the growing new field of nanotechnology concerns methods of manufacturing materials, artefacts, machines, and even robots on extremely small scale.[23] Nanobots are simple robots, but built on an extremely small scale. Unlike a large robot, a nanobot's programming cannot be highly sophisticated. But if you have hundreds, thousands, or even millions of nanobots, then the possibilities are very rich indeed. Like an ant colony, the rules each nanobot obeys can be very simple, and yet useful results could emerge from the activity of an entire swarm.[24]

In manufacturing, tiny machines could assemble materials and even whole products with properties unattainable in any conventional way. Another application lies in purification and extraction. Like ants running amok, tiny nanobots could work through (say) waste, extracting particles of metal for recycling or removing contaminants and impurities from mixtures. Imagine taking a cup of sand and a cup of seeds (each the same size as a grain of sand) and mixing them up. Separating them again is a slow and tedious process. However, a swarm of robot ants could sort them without any difficulty. A medical application of this idea would be to inject swarms of nanobots into a patient, providing on-the-spot detail of illness, helping to repair injuries, and helping the body's natural defences by capturing and removing viruses.

Finally another, more distant prospect is self-replicating robots: machines that duplicate themselves using local raw materials. There are many advantages to machines that can duplicate themselves. For one thing, it could simplify the problem of manufacturing and transporting them.[25] However, self-replication also carries with it the potential for mayhem. As society has already learned to its cost, self-replicating software leads to computer viruses. So self-replicating nanobots might mimic plagues of locusts, breeding uncontrollably and stripping resources from their environment, whatever that environment happened to be. Drexler[26] popularized the term "grey goo" for catastrophic swarms of nanobots destroying the world by eating away component materials of the natural and human environment. As we saw in the previous chapter, the release of organisms into new environments has already led to major environmental problems, even without nanotechnology.

The potential benefits from all the above developments are huge, but so too is the potential for unwelcome side effects. More generally, the sheer pace of technological innovation has become a major driver of social change. As we saw in Chap. 12, the rate at which new technologies are being introduced is accelerating (Fig. 12.1). So-called *transhumanists* see this trend as a positive sign. They speculate that advances in computer technology will eventually lead to a *technological singularity*. This would occur when artificial intelligence exceeds human intelligence. At that point, technology would not only transform society, but could also radically alter human nature and the course of future evolution.[27] Note, however, that the argument assumes that technology can continue to accelerate as it has in the recent past.

## The Big Picture

For most of this book, I have concentrated on danger; danger that lurks as unexpected side effects of things we do. This is not merely negative thinking. Almost by definition, unintended side effects are not what people want, so almost inevitably, they are negative. Nevertheless, not all unplanned trends are negative. Some are neutral; some are actually positive. If there are trends we want to encourage, then the clever way to make them happen is to arrange things so that the trends emerge naturally as side effects of things people do anyway. This approach is already widely used. Governments tweak interest rates to change investment patterns. Let's conclude by looking at some examples of consciously driven trends that are making things better.

When society is determined enough, progress can be made. We saw a good example in Chap. 11. The aeronautics industry has worked to make air travel safer. An even better example is the way some communities have been able to reduce the death toll on their roads.

The post-war history of road fatalities in many western countries tells a story very similar to that for air travel. Increases in the number of cars on roads during the post war period led to dramatic increases in the numbers of road fatalities. The first community to take serious action to reverse the trend was the Australian state of Victoria. By 1968, the state's annual death toll from road accidents had risen to more than 1,300 people per year. This was far from the world's highest annual death toll, but represented an unacceptably high proportion of travellers. It is even more alarming when you consider that deaths are only a fraction of the total numbers of people injured on roads each year.

Reporting the numbers of deaths failed to stem the carnage on the roads, so the government decided to take more direct action. In 1970, the Victorian government became the first in the world to pass legislation making the wearing of seatbelts compulsory. The reduction in road deaths was immediate (Fig. 20.1).[28] Other governments soon followed suit. Over the years a series of further measures and public awareness campaigns led to a steady decrease in the number of deaths per

**Fig. 20.1** Reversal of a rapidly increasing trend in numbers of road deaths *per annum* for the city of Melbourne following the introduction in seat belt laws in 1970[28]

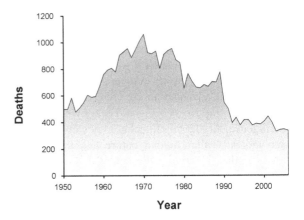

annum. By 2013, the campaign had reduced the road toll to less than 300 deaths per year, about a quarter of what it had been, despite there being more than twice as many cars on the roads.

Ultimately, society cannot protect everyone from every possible danger. The problem is one of balance. Life is risky. Almost everything we do involves some degree of risk. Sit down to eat a meal and there is a chance you might choke on a mouthful of food. Step into a bath and there is a chance you might slip and break your neck.

As we saw earlier, many situations involve a trade-off. Trade-offs are inevitable when you have two or more competing goals. If you plan to hold a party, you want it to be safe, but on the other hand you do not want to be crippled by so many safety regulations that you cannot hold the party at all. In the end you have to strike a balance: you take reasonable care, but there is always a risk of something going wrong. Many of the world's great issues are like that. There may be a straightforward solution to one problem, but sometimes the price is that we create an even bigger problem elsewhere.

Nevertheless, if society fails to identify problems, if it fails to set goals, then unintended trends can drift rapidly into crisis, as we saw in the previous chapter. Fortunately, the world community is already trying to create trends to address major problems. In the year 2000, the United Nations held a Millennium Summit in New York. Leaders of 189 countries who attended signed the UN Millennium Declaration, which led to the following set of Millennium Development Goals.[29]

Goal 1:   Eradicate extreme poverty and hunger
Goal 2:   Achieve universal primary education
Goal 3:   Promote gender equality and empower women
Goal 4:   Reduce child mortality
Goal 5:   Improve maternal health
Goal 6:   Combat HIV/AIDS, malaria and other diseases
Goal 7:   Ensure environmental sustainability
Goal 8:   Develop a global partnership for development

**Table 20.1** Sample of United Nations world development indicators

| Indicator | 1990 | 2010 |
|---|---|---|
| People below minimum level of dietary energy consumption (%) | 18.6 | 12.5 |
| Percentage enrolled in primary education (%) | 81.9 | 91.2 |
| Proportion of pupils completing primary school (%) | 80.5 | 90.6 |
| 15–24 years who can both read and write (%) | 83.4 | 89.5 |
| Seats held by women in national parliament (%) | 12.8 | 20.8 |
| Under-5 mortality rate (per 1,000 live births) | 87 | 50 |
| Infant mortality rate (per 1,000 live births) | 61 | 36 |
| 1-year-old children immunized against measles (%) | 72 | 84 |
| Maternal mortality (per 100,000 live births) | 400 | 210 |
| Contraceptive prevalence rate (women aged 15–49) (%) | 55 | 63.2 |
| Adolescent birth rate (per 1,000 women aged 15–19) | 59.3 | 48.6 |
| HIV incidence (per 100 people aged 15–49) | 0.08 | 0.06 |
| Deaths from tuberculosis (per 100,000 population) | 24 | 14 |
| People using an improved drinking water (%) | 76 | 89 |
| People using improved sanitation (%) | 49 | 64 |
| Proportion of urban population living in slums (%) | 46.2 | 32.7 |
| Proportion of land area covered by forest (%) | 32 | 31 |
| Carbon dioxide emissions (millions of metric tons) | 21,550 | 31,387 |
| Consumption of ozone-depleting substances | 1,121,310 | 31,837 |
| Proportion of fish stocks overexploited (%) | 18.6 | 29.9 |
| Proportion of terrestrial and marine areas protected (%) | 8.3 | 14.0 |
| Species considered safe from extinction (%) | 92.1 | 91.3 |

*Source* United Nations (2013). Note that some figures are based on 1991–1992, or 2011

Set against these goals were 21 targets, with 60 measurable indicators to monitor progress (Table 20.1). As the table shows, in 2010, the world was making progress on almost every human and social goal, but was going backwards on most of the environmental indicators. However, overall progress on a single indicator hides a mixed picture and highlights the danger of reducing complex problems to simple metrics. On many indicators there is considerable variation, both between regions and between countries. For some indicators, improvements in a few countries (e.g. population control in China) mask lack of progress elsewhere.

As we have seen, the sheer complexity of the world around us means that unexpected side effects are always possible. However, in most cases, the side effects are not so obscure, and not so unpredictable. The real problem is failure, or unwillingness, to consider the bigger picture. In so many cases, the issue boils down to traits, such as the desire for status, that we inherit from our animal past.

Information technology is part of the problem. But it is also part of the solution. For every problem IT creates there is also a problem that it solves. The same technology that helps companies expand into global corporations also helps dispersed minority groups and cultures to maintain their identity. Information Technology (IT) provides a means to address complex global issues. For example IT makes it possible to identify and monitor global biodiversity and environments and help conserve the planet's living resources. It provides the tools to carry out virtual experiments to test the consequences of plans and actions ahead of time.[30] Global communication helps to break down barriers by increasing awareness and understanding of other countries and cultures.

Like the ants, side effects of our daily activity will continue to create unplanned and unexpected social outcomes. Unlike the ants we humans have the ability to plan. We can, if we care to, look ahead and see the likely consequences of our actions. The real question is whether we can restrain the drives we inherit from our ancient past. Can we put the world's interests ahead of individual or national desires? Or will we go on living in denial, inviting one calamity after another? Can we learn the lessons of history? Can we adapt?

## End Notes

[1]Ernest Hemingway, opening the final chapter of *Death in the Afternoon.*
[2]See Chap. 9.
[3]See Chap. 2.
[4]See Chap. 1.
[5]Ralston Saul (1997).
[6]In 2012 a reference list compiled for the Australian Research Council listed 22,413 research journals. A high proportion of these could be considered as science journals. This number had grown by more than 2000 titles from the previous list, compiled in 2010. A similar compilation of conferences listed nearly 2000 regular meetings.
  http://www.arc.gov.au/era/era_2012/era_2012.htm.
[7]Cf. Chap. 11.
[8]For example, a recent survey found that about 1/3 of all Australians did not even know that the Earth takes a year to revolve around the sun. See http://www.science.org.au/reports/science-literacy.html. Downloaded 1 August 2013.
[9]Henderson (2012).
[10]Megaloganis (2010).
[11]Pariser (2012).
[12]See for instance, the discussion of early Web sites in Chap. 12.
[13]UNEP (1992). The Convention on Biological Diversity. United Nations Environment Programme, Rio de Janeiro. www.cbd.int/convention/text/.
[14]Source: http://web.ornl.gov/sci/techresources/Human_Genome/ (downloaded 15 September 2013).

[15]See for instance the U.S. National Institute of Health ENCODE Project. http:// www.genome.gov/10005107 (downloaded 15 September 2013).

[16]The technique, known as the Polymerase Chain Reaction (PCR), can generate millions of copies of a gene. It is a crucial method used in gene cloning and in studies to determine DNA sequences.

[17]Simon et al. (1983).

[18]Isaacs et al. (2011).

[19]"Life on track for new genetic code." *New Scientist* 23 July 2011, p. 8.

[20]The Association for the Advancement of Artificial Intelligence (AAAI) provides links to starting points about many topics in current AI research and application. Source: www.aaai.org (downloaded 17 September 2013).

[21]Christopher Steiner (2012). Knight Capital's algorithmic fiasco won't be the last of its kind. *Forbes*. http://www.forbes.com/sites/christophersteiner/2012/08/02/ knight-capitals-algorithmic-fiasco-wont-be-the-last-of-its-kind/.

[22]See for instance http://www.robocup.org/ (downloaded 2 August 2013).

[23]A nanometre is one billionth of a metre. At the extreme nanotechnology deals with manufacturing materials and objects at the atomic scale. However the term is often used loosely to refer to the trend towards manufacturing anything extremely small.

[24]For an introduction to swarms see, for example, Sadedin and Duéñez-Guzmán (2012).

[25]Self-reproducing machines would still need building blocks. So either they would need to build copies from prefabricated components or else to grow by retrieving and processing raw materials.

[26]Drexler (1986).

[27]See for instance Ulam (1958) and Kurzweil (2005). Calculations by various authors have put the singularity occurring before 2150.

[28]Source: Road Safety Victoria. www.roadsafety.vic.gov.au/facts (accessed 15 September 2013).

[29]United Nations (2013).

[30]Green (2004), Chap. 11.

# References

ABS (2009). Australian social trends. Australian Bureau of Statistics. www.abs.gov.au/, Downloaded July 22, 2013.

Albert, R., Jeong, H., & Barabasi, A.-L. (1999). Diameter of the World-Wide Web. *Nature, 401*, 130–131.

Alvarez, L. W., Alvarez, W., Asaro, F., & Michel, H. V. (1980). Extraterrestrial cause for the Cretaceous-Tertiary extinction. *Science, 208*, 1095–1108.

American Academy of Pediatrics. (2006). Committee on communications policy statement: children, adolescents, and advertising. *Pediatrics, 118*(6), 2563–2569.

Anderson, S., & Cavanagh, J. (2000). *The Top 200: The rise of Global Corporate power.* Washington: Institute for Policy Studies.

Anonymous (2006). Dying alone. *The Daily Telegraph*, Sydney, February 22, 2006.

Ardrey, R. (1966). *The territorial imperative: A personal inquiry into the animal origins of property and nations.* London: Collins.

Ballantyne, A. P., Alden, C. B., Miller, J. B., Tans, P. P., & White, J. W. C. (2012). Increase in observed net carbon dioxide uptake by land and oceans during the past 50 years. *Nature, 488*, 70–72. doi:10.1038/nature11299.

Barabási, A.-L., & Albert, R. (1999). Emergence of scaling in random networks. *Science, 286*(5439), 509–512.

Bearman, P. S., Moody, J., & Stovel, K. (2004). Chains of affection: The structure of adolescent romantic and sexual networks. *American Journal of Sociology, 110*(1), 44–91.

Bilalić, M., Smallbone, K., McLeod, P., & Gobet, F. (2009). Why are (the best) women so good at chess? Participation rates and gender differences in intellectual domains. *Proceedings of the Royal Society B, 276*(1659), 1161–1165. doi:10.1098/rspb.2008.1576.

Bilofsky, H. S., & Burks, C. (1988). The GenBank genetic sequence data bank. *Nucleic Acids Research, 16*(5), 1861–1863.

Bonabeau, E., Dorigo, M., & Theraulaz, G. (1999). *Swarm intelligence: From natural to artificial systems.* Oxford: Oxford University Press.

Bossomaier, T. R. J., & Green, D. G. (1998). *Patterns in the sand.* Sydney: Allen and Unwin.

Boudreau, K. J., & Lakhani, K. R. (2013, April). Using the crowd as an innovation partner. *Harvard Business Review*, 61–69. HBR Reprint R1304C.

Bell, G., & Gemmell, J. (2009). *Total recall—How the e-memory revolution will change everything* (288pp.). New York: Dutton (Penguin Group). ISBN 978-0-525-95134-6.

Block, J. (1982). Assimilation, accommodation and the dynamics of personality development. *Child Development, 53*, 281–295.

Bowles, S. (2011a). Cultivation of cereals by the first farmers was not more productive than foraging. *Proceedings of the National Academy of Science, 108*(12), 4760–4765. doi:10.1073/pnas.1010733108.

Bowles, S. (2011b). Better mousetraps? Beware simple history. *New Scientist, 211*(2823), 26–27.

D. G. Green, *Of Ants and Men*, DOI: 10.1007/978-3-642-55230-4,
© Springer-Verlag Berlin Heidelberg 2014

Bransden, T. G., & Green, D. G. (2005). Getting along with your neighbours-emergent cooperation in networks of adaptive agents. In A. Ohuchi, K. Suzuki, M. Gen & D. G. Green (Eds.), *Workshop on Intelligent and Evolutionary Systems (IES2005)*. Japan: Future University-Hakodate. www.waseda.jp/sem-genlab/ ~ IES2005/

Bruggink, G. M. (2000, August). Remembering Tenerife. *Air Line Pilot*, p. 18.

Caporael, L. R. (1976). Ergotism: The Satan loosed in Salem? *Science, 192*(4234), 21–26.

Cheong, M., Ray, S., & Green, D. G. (2012). Large-scale socio-demographic pattern discovery on microblog metadata. In *12th International Conference on Intelligent Systems Design and Applications (ISDA)* (pp. 909–914). IEEE.

Churchill, W. S. (1923). *The World Crisis 1911–1918*. Republished by Simon & Schuster in 2005. ISBN 0743283430

Cole, H. S. D., Freeman, C., Jahoda, M., & Pavitt, K. L. R. (Eds.) (1973). *Thinking about the future—A critique of the limits to growth*. London: Chatto and Windus, London for Sussex University Press. ISBN 0-85621-018-8.

Costanza, R., Cumberland, J. H., Daly, H., Goodland, R., & Norgaard, R. B. (2007). *An introduction to ecological economics*. Available as an e-book from Encyclopedia of Earth. www.eoearth.org/view/article/150045, Originally published 1997 by St. Lucie Press and International Society for Ecological Economics.

Dahl, O.-J., & Nygaard, K. (1966). SIMULA—An ALGOL-based simulation language. *Communications of Association for Computer Machinery, 9*(9), 671–678.

Daly, H., & Farley, J. (2010). *Ecological economics: Principles and applications*. New York: Island Press. ISBN 1597266817.

Davies, R. (2010). Marking the 50th anniversary of the Platt report: From exclusion, to toleration and parental participation in the care of the hospitalized child. *Journal of Child Health Care, 14*(1), 6–23.

Deane, P. (1965). *The first industrial revolution*. New York: Cambridge University Press.

Diamond, J. (2005). *Collapse—How societies choose to succeed or fail*. New York: Viking. ISBN 0-670-03337-5.

Diethelm, P., & McKee, M. (2009). Denialism: What is it and how should scientists respond? *European Journal of Public Health, 19*(1), 2–4.

de Graaf, J., Wann, D., & Naylor, T. N. (2001). *Affluenza: The all-consuming epidemic*. San Francisco: Berrett-Koehler. ISBN 1-57675-199-6.

Drexler, E. (1986). *Engines of creation—The coming era of nanotechnology*. New York: Anchor Books.

Dronkers (2003). Has the Dutch nobility retained its social relevance during the 20th century? *European Sociological Review, 19*, 81.

Dunbar, R. (1996). *Grooming, gossip and the evolution of language*. London: Faber and Faber.

Eldredge, N., & Gould, S. J. (1972). Punctuated equilibria: an alternative to phyletic gradualism. *Models in Paleobiology, 82*, 115.

Epley, N., Waytz, A., Akalis, S., & Cacioppo, J. T. (2008). When we need a human: motivational determinants of anthropomorphism. *Social Cognition, 26*(2), 143–155.

Erdős, P., & Rényi, A. (1959). On random graphs. I. *Publicationes Mathematicae, 6*, 290–297.

ESRL (2013). *Trends in atmospheric carbon dioxide*. Earth System Research Laboratory. www.esrl.noaa.gov/gmd/ccgg/trends/global.html, Accessed October 28, 2013.

Fancher, R. E. (1985). *The intelligence men: makers of the IQ controversy*. New York: Norton & Co.

Fang, I. (1997). *A history of mass communication*. Boston: Focal Press. ISBN 0-240-80254-3.

FAO (2011). *Land Area*. United Nations Food and Agricultural Organization. http://data.un.org

Fenner, F., Henderson, D. A., Arita, I., Ježek, Z., & Ladnyi, I. D. (1988). Smallpox and its eradication. World Health Organization, Geneva. ISBN 92 4 156110 6. http://whqlibdoc.who.int/smallpox/9241561106.pdf, Downloaded July 31, 2013.

Forrester, J. W. (1971). *World dynamics*. Cambridge: Wright-Allen Press.

Fox, D. (2010). In our own image. *New Scientist, 2788*, 32–37.

Freedman, L. (2005). *The official history of the Falklands campaign* (806 pp.). Volume II War and diplomacy. Oxford: Routledge. ISBN 0-7146-5207-5

Gardner, D. (2010). *Future babble: Why expert predictions fail-and why we believe them anyway.* Toronto: McClelland and Stewart Ltd. ISBN 978-0-7710-3519-7.

Gehl, J. (2010). *Cities for people* (285 pp.). New York: Island Press. ISBN 9781597265737

Gerhardt, S. (2010). *The selfish society—How we all forgot to love one another and made money instead* (388 pp.). London: Simon and Schuster. ISBN 978-1-84737-571-1

Gero, D. (2003). *Aviation disasters: The World's Major Civil Airliner crashes since 1950.* Sutton.

Gillan (2006). Body of woman, 40, lay unmissed in flat for more than two years. *The Guardian,* April 14, 2006. http://www.guardian.co.uk/uk/2006/apr/14/audreygillan.uknews2.

Gleick, J. (1999). *Faster—The speeding up of almost everything.* New York: Pantheon. ISBN 9780679408376.

Goodall, J. (1999). *Reason for hope: A spiritual journey.* New York: Warner Books.

Gould, S. J. (1997). *Full house: The spread of excellence from Plato to Darwin.* New York: Harmony Books. ISBN 0-517-70394-7.

Grattan, J. (2006). Aspects of Armageddon: An exploration of the role of volcanic eruptions in human history and civilization. *Quaternary International, 151*(1), 10–18.

Green, D. G. (1982). Fire and stability in the postglacial forests of southwest Nova Scotia. *Journal of Biogeography, 9,* 29–40.

Green, D. G. (2004). *The serendipity machine.* Sydney: Allen and Unwin.

Green, D. G., Klomp, N. I., Rimmington, G. R., & Sadedin, S. (2006). *Complexity in landscape ecology.* Amsterdam: Springer.

Green, D. G., Leishman, T. G., & Sadedin, S. (2006b). The emergence of social consensus in simulation studies with Boolean networks. In S. Takahashi, D. Sallach & J. Rouchier (Eds.), *Proceedings of the first world congress on social simulation* (Vol. 2, pp. 1–8), Kyoto.

Green, D. G., Leishman, T. G., & Sadedin, S. (2006c). Dual phase evolution—A mechanism for self-organization in complex systems. *InterJournal,* 1–8. Cambridge: New England Complex Systems Institute. ISSN 1081-0625

Green, D. G., Liu, J., & Abbass, H. (2014). *Dual phase evolution: From theory to practice.* Berlin: Springer. ISBN-13 978-1441984227, 978-1-4419-8423-4 (eBook)

Gregg, J. (2013). *Are Dolphins really smart? The Mammal behind the myth.* Oxford: Oxford University Press. ISBN 978-0-19-966045-2.

Hamilton, C., & Denniss, R. (2005). *Affluenza: When too much is never enough* (224 pp.). Crow's Nest NSW Australia: Allen and Unwin. ISBN 978-1-74114-671-4.

Hardin, G. (1968). The tragedy of the commons. *Science,162*(3859), 1243–1248. doi:10.1126/science.162.3859.1243.

Harkin, J. (2009). *Cyburbia* (274 pp.). London: Little Brown. ISBN 978-1-408-70113-3

Harley, H. E., Erika, A., Putman, E. A., & Roitblat, H. L. (2003). Bottlenose dolphins perceive object features through echolocation. *Nature, 424,* 667–668.

Hayek, F. A. (1982). *Law, legislation and liberty.* London: Routledge.

Henderson, M. (2012). *The geek manifesto: Why science matters.* London: Random House.

Herman, L. M., Pack, A. A., & Morrel-Samuels, P. (1993). Representational and conceptual skills of dolphins. In H. L. Roitblat, L. M. Herman & P. E. Nachtigall (Eds.) *Language and communication: Comparative perspectives. Comparative cognition and neuroscience* (pp. 403–442). Hillsdale: Lawrence Erlbaum Associates.

Hoffman, P. (1988). *The man who loved only numbers: The story of Paul Erdos and the search for mathematical truth.* New York: Hyperion. ISBN 0-7868-6362-5.

Hogeweg, P., & Hesper, B. (1983). The ontogeny of the interaction structure in bumblebee colonies: A MIRROR model. *Behavioral Ecology and Sociobiology, 12,* 271–283.

Holmes, T. H., & Rahe, R. H. (1967). The social readjustment rating scale. *Journal of Psychosometric Research, 11*(2), 213–218.

Holzmann, G. J., & Pehrson, B. (1994). The first data networks. *Scientific American, 270*(1), 124–129.

Horne, A. (1962). *The price of glory: Verdun 1916.* London: Macmillan.

Hoofnagle, M. (2007). Denialism Blog. http://scienceblogs.com/denialism/about/

Huffman, M. A., Nahallage, C. A. D., & Leca, J.-B. (2008). Cultured monkeys: Social learning cast in stones. *Current Directions in Psychological Science, 17*(6), 410. doi:10.1111/j.1467-8721.2008.00616.x.

Isaacs, F. J., Carr, P. A., Wang, H. H., Lajoie, M. J., Sterling, B., Kraal, L., et al. (2011). Precise manipulation of chromosomes in vivo enables genome-wide codon replacement. *Science, 333*(6040), 348–353. doi:10.1126/science.1205822.

ISOC (2012). *A brief history of the internet.* The Internet Society. www.internetsociety.org.

Jackson, T. (2008). What politicians dare not say. *New Scientist, 2678,* 42–43 (Articles in NS pp. 40–54).

Jain, H. K. (2010). *Green revolution: History, impact and future.* Houston: Studium Press. ISBN 9781933699639.

Jani, V. M. (2000). Whistle matching in wild bottlenose dolphins (*Tursiops truncatus*). *Science, 289,* 1355–1357.

Jones, T. B., & Kamil, A. C. (1973). Tool-making and tool-using in the northern blue jay. *Science, 180,* 1076–1078.

Kasser, T. (2002). *The high price of materialism.* Cambridge: MIT Press.

Keys, D. (2000). Catastrophe: An investigation into the origins of the modern world. New York: Ballantine Books.

Kingsford, R. T. (2000). Ecological impacts of dams, water diversions and river management on floodplain wetlands in Australia. *Australian Ecology, 25,* 109.

Klayman, J., & Young-won, H. (1987). Confirmation, disconfirmation, and information in hypothesis testing. *Psychological Review, 94*(2), 211–228. doi:10.1037/0033-295X.94.2.211.

Klinenberg, E. (2012). Living alone is now the norm. *Time Magazine, 179*(10), 38–41.

Koestler, A. (1964). *The act of creation.* New York: Macmillan.

Korb, K., & Nicholson, A. (2011). *Bayesian artificial intelligence* (2nd ed.). Boca Raton: CRC Press.

Kübler-Ross, E. (1973). *On death and dying.* New York: Macmillan.

Kurzweil, R. (2005). *The singularity is near.* New York: Viking Press. ISBN 0-670-03384-7.

Laird (2007). *PULL: Networking and Success since Benjamin Franklin* (pp. 439). Cambridge: Harvard University Press.

Larman, C. (2005). *Applying UML and patterns—Introduction to OOA/D & iterative development* (3rd ed.). Englewood Cliffs: Prentice Hall PTR.

Lebow, V. (1955). Price competition in 1955. *Journal of Retailing, 31*(1), 5–10.

Lever, C. (2001). *The cane toad. The history and ecology of a successful colonist.* West Yorkshire: Westbury Academic and Scientific Publishing.

Levy, D. H., Shoemaker, E. M., & Shoemaker, C. S. (1995). Comet Shoemaker-Levy 9 meets Jupiter. *Scientific American, 273*(2), 68–75.

Lloyd, J. (2004). What the media are doing to our politics (218 pp.). London: Constable & Robinson Ltd. ISBN 1-84119-900-1.

Mace, R. (2009). On becoming modern. *Science, 324*(5932), 1280–1281. doi:10.1126/science.1175383.

Mackay, C. (1841). *Extraordinary popular delusions and the Madness of crowds* (740 pp.). London: Richard Bentley. Reprinted in A. Tobias (Ed.) (1980). New York: Three Rivers Press. ISBN 0-517-88433-X.

Macellini, S., Maranesi, M., Bonini, L., Simone, L., Rozzi, S., Ferrari, P. F., et al. (2012). Individual and social learning processes involved in the acquisition and generalization of tool use in macaques. *Philosophical Transactions of the Royal Society B: Biological Sciences, 367*(1585), 24–36.

Mander, J. (1999). Economic globalization: the era of corporate rule. 19th Annual E.F. Schumacher Lectures. *New Economics Institute.* http://neweconomy.net/publications/lectures/, Accessed April 21, 2014.

Marks, P. (2009). NASA criticised for sticking to imperial units. *New Scientist*. www. newscientist.com/article/dn17350-nasa-criticised-for-sticking-to-imperial-units.html, Downloaded July 20, 2013.

McKenzie, B., & Rapino, M. (2011). Commuting in the United States: 2009. American Community Survey Reports, ACS-15. US Census Bureau. http://www.census.gov/prod/2011pubs/acs-15.pdf, Downloaded July 29, 2013.

Meadows, D. H., Meadows, D. L., Randers, J., & Behrens, W. W. (1972). *The limits to growth*. New York: Universe Books. ISBN 0-87663-165-0.

Megaloganis, G. (2010). *Quarterly Essay 40 Trivial pursuit: Leadership and the end of the reform era*. Melbourne: Black Inc. ISBN 9781863954983. http://www.readings.com.au/product/9781863954983/george-megalogenis-quarterly-essay-40-trivial-pursuit-leadership-and-the-end-of-the-reform-era

Merton, R. K. (1936). The unanticipated consequences of purposive social action. *American Sociological Review, 1*(6), 894–904.

Milgram, S. (1967). The small-world problem. *Psychology Today, 1*, 60–67.

Miller, G. A. (1956). The magical number seven, plus or minus two: Some limits on our capacity for processing information. *Psychological Review, 63*(2), 81–97.

Morris, D. (1967). *The naked ape: A zoologist's study of the human animal*. New York: McGraw Hill. ISBN 0070431744.

Morris, D. R. (1965). *The washing of the spears: The rise and fall of the Zulu Nation*. Pimlico. ISBN 978-0-712-66105-8.

Nagurney, A. (2006). *Supply Chain Network Economics: Dynamics of Prices, Flows, and Profits*. Northampton: Edward Elgar Publishing. ISBN 1-84542-916-8.

Netcraft (2012). http://news.netcraft.com/archives/2012/03/05/march-2012-web-server-survey.html

Nickerson, R. S. (1998). Confirmation bias: A ubiquitous phenomenon in many guises. *Review of General Psychology, 2*(2), 175–220.

Paperin, G., Green, D. G., & Sadedin, S. (2011). Dual phase evolution in complex adaptive systems. *Journal of the Royal Society Interface, 8*(58), 609–629. doi:10.1098/rsif.2010.0719.

Pearce, J. (2001). *Small is still beautiful*. London: Harper Collins. ISBN 0-00-714215-3.

Penn, M. J., & Zalesne, E. K. (2007). *Microtrends—The small forces behind today's big changes* (425 pp.). New York: Allen Lane (Penguin). ISBN 978-1-846-14062-4.

Piaget, J. (1971). *Biology and knowledge* (384 pp.). Edinburgh: Edinburgh University Press.

Pinker, S. (2011). *The better angels of our nature—Why violence has declined* (832 pp.). New York: Viking. ISBN 978-0670022953.

Ponting, C. (1991). *A green history of the world*. London: Penguin Books. ISBN 0-14-01-6642-4.

Poscente, V. (2008). *The age of speed: Learning to thrive in a more-faster-now world*. New York: Ballantine Books.

Pringle, H. (2013). The origins of creativity. *Scientific American, 388*(2), 24–29.

Putnam, R. (2001). *Bowling alone: The collapse and revival of american community*. London: Simon & Schuster.

Reeve, T. (20110). How many cameras in the UK? Only 1.85 million, claims ACPO lead on CCTV. *Security News Desk*. www.securitynewsdesk.com/2011/03/01/how-many-cctv-cameras-in-the-uk/

Robertson, J. (1958). *Young children in hospital*. London: Harper and Rowe.

Rudel, T. K., & Hooper, L. (2005). Is the pace of social change accelerating? Latecomers, common languages, and rapid historical declines in fertility. *International Journal of Comparative Sociology, 46*(4), 275–296. doi:10.1177/0020715205059204.

Sadedin, S., & Duéñez-Guzmán, E. A. (2012). Harnessing the swarm: Technological applications of collective intelligence. In A. Poiani (Ed.), *Pragmatic evolution: Applications of evolutionary theory* (pp. 234–258). Cambridge: Cambridge University Press. ISBN-13 9780521760553.

Ralston Saul, J. (1997). *The unconscious civilization*. Ringwood: Penguin Books. ISBN 0-14-036464-7.

Raymond, R. (1986). *Starfish wars—Coral death and the crown-of-thorns*. South Melbourne: Macmillan.

Schechter, B. (1998). *My brain is open: The mathematical journeys of Paul Erdos*. New York: Simon and Schuster. ISBN 0-684-84635-7.

Schumacher, E. F. (1973). *Small is beautiful—A study of economics as if people mattered*. London: Vintage Books. ISBN 0-09-922561-1.

Powell, A., Shennan, S., & Thomas, M. G. (2009). Late pleistocene demography and the appearance of modern human behavior. *Science, 324*(5932), 1298–1301. doi:10.1126/science.1170165.

Shermer, M. (2010). Living in denial: When a sceptic isn't a sceptic. *New Scientist*. www.newscientist.com/special/living-in-denial, May 18, 2010.

Shlain, L. (1998). *The alphabet versus the goddess: The conflict between word and image* (464 pp.). Harmondsworth: Viking Penguin. ISBN 0-670-87883-9.

Simon, R., Priefer, U., & Pühler, A. (1983). A broad host range mobilization system for in vivo genetic engineering: transposon mutagenesis in gram negative bacteria. *Nature Biotechnology, 1*(9), 784–791.

Small, M. (2004). The natural history of children. In S. Olfman (Ed.), *Childhood lost: How American culture is failing our kids* (pp. 3–18). Westport: Praeger.

Smith, A. (1776). *An inquiry into the nature and causes of the wealth of nations*. London: W. Strahan and T. Cadell (Reprinted many times by other publishers).

Soper, K. (2008). The good life. *New Scientist, 2678*, 54.

Spennemann, D. (2005). *Digital micronesia-an electronic library & archive*. Albury: Charles Sturt University. http://marshall.csu.edu.au/

Stocker, R., Green, D. G., & Newth, D. (2001). Consensus and cohesion in simulated social networks. *Journal of Artificial Societies and Social Simulation, 4*. http://www.soc.surrey.ac.uk/JASSS/4/4/5.html

Stocker, R., Cornforth, D., & Green, D. G. (2003). A simulation of the impact of media on social cohesion. *Advances in Complex Systems, 6*(3), 349–359. http://www.worldscinet.com/acs/06/0603/S0219525903000931.html

Surowiecki, J. (2004). *The wisdom of crowds*. New York: Anchor Books.

Suter, K. (2000). *In defence of globalisation*. Sydney: UNSW Press.

Suzuki, D., & Taylor, D. R. (2009). *The big picture*. Sydney: Allen and Unwin.

Taleb, N. N. (2007). *The black swan—The impact of the highly improbable*. New York: Random House. ISBN 978-1-4000-6351-2.

Taylor, A. J. P. (1963). *The First World War: An illustrated history*. London: Penguin Books. ISBN 0-14-002481-6.

Taylor, R., & Morrissey, K. (2004). Coping with pollution: dealing with waste. In F. Harris (Ed.) *Global environmental issues* (pp. 229–264). New York: Wiley. ISBN 0-470-84561-9.

Taverne, R. (2005). *The March of unreason—Science, democracy and the new fundamentalism* (310 pp.). Oxford: Oxford University Press. ISBN 0-19-280485-5.

Tenner, E. (1996). *Why things bite back: Technology and the revenge of unintended consequences*. Alfred A. Knopf. ISBN 0-679-42563-2.

Tetlock, P. E. (2005). *Expert political judgment: How good is it? How can we know?* Princeton: Princeton University Press.

Thomas, G., & Witts, M. (2011). *The day the world ended: Mont Pelee Earthquake 1902*. New York: Premier Digital Publishing. ISBN 9781937624002.

Toffler, A. (1970). *Future shock*. New York: Random House.

Ulam, S. (1958). Tribute to John von Neumann. *Bulletin of the American Mathematical Society, 64*(3/2), 5.

United Nations (2013). *Millennium development goals indicators*. United Nations Statistics Division. http://mdgs.un.org/unsd/mdg/, Downloaded October 12, 2013.

United States Census Bureau (2012). *The 2012 statistical abstract.*www.census.gov/compendia/statab/, Downloaded December 13, 2013.

Veblen, T. (1904). *The theory of the business enterprise.* New Brunswick: Transaction Books. ISBN 0-87855-699-0.

Vornovytskyy, M., Gottschalck, A., & Smith, A. (2012). Household Debt in the U.S.: 2000 to 2011. United States Census. www.census.gov/people/wealth/files/Debt%20Highlights2011.pdf, Downloaded December 13, 2013.

Walters, J. M. (2000). *Aircraft accident analysis: Final reports.* New York: McGraw-Hill Professional.

Waytz, A., Epley, N., & Cacioppo, J. T. (2010). Social cognition unbound: Insights into anthropomorphism and dehumanization. *Current Directions in Psychological Science, 19,* 58–62.

Weizenbaum, J. (1966). ELIZA—A computer program for the study of natural language communication between man and machine. *Communications of the ACM, 9*(1), 36–45. doi:10.1145/365153.365168.

Whiston, T. G. (1991). Forecasting the world's problems: The Last Empire: The corporatization of society and diminution of self. *Futures, 23*(2), 163–178.

White, T. D., Asfaw, B., Beyene, Y., et al. (2009). *Ardipithecus ramidus* and the paleobiology of early hominids. *Science, 326*(5949), 75–86. doi:10.1126/science.1175802.

Wiener, N. (1948). *Cybernetics or control and communication in the animal and the machine.* Oxford: Wiley

Williams, T. M. (1986). *The impact of television—A natural experiment in three communities.* New York: Academic Press.

Woolf, A. (2000). Witchcraft or mycotoxin? The Salem witch trials. *Clinical Toxicology, 38*(4), 457–460.

Warner, J. (2013). *The legend of the Forty-Seven Ronin: A history of one of the greatest samurai stories of all time.* London: BookCaps Study Guides. ISBN 9781621075110.

Washington, H., & Cook, J. (2011). *Climate change denial: Heads in the sand.* New York: Earthscan. ISBN 978-1-84971-335-1.

Watts, D. J., & Strogatz, S. H. (1998). Collective dynamics of 'small-world' networks. *Nature, 393,* 440–442.

Wilson, E. O. (1975). *Sociobiology—The new synthesis.* Cambridge: Harvard University Press.

Wiseman, R. (2003). *The luck factor.* London: Random House.

Wolfe, T. (1979). *The right stuff.* New York: Bantam.

Wright, R. (2005). *A short history of progress.* Melbourne: Text Publishing Co. ISBN 1-920885-79-X.

# Index